Finite Elements Using Natural Strains & Basis Transformation

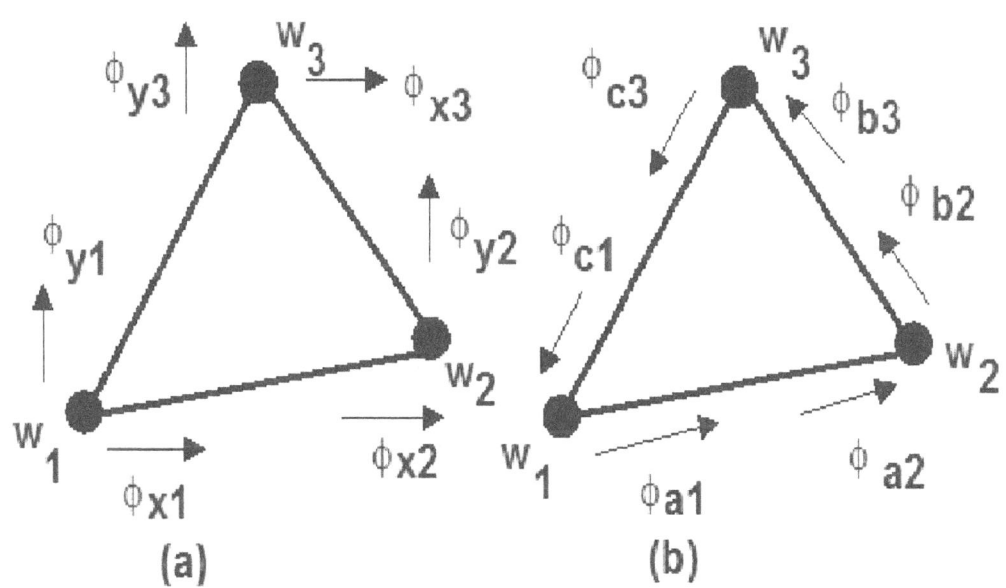

(a) (b)

G. A. MOHR, PhD
WORLD HONS MULT

Finite Elements Using Natural Strains & Basis Transformation

G. A. MOHR, PHD
WORLD HONS MULT.

© G. A. Mohr, 2019

G. A. Mohr
Finite Elements Using Natural Strains & Basis Transformation

TRI

Transworld Research & Innovation
9 Hampstead Drive
Hoppers Crossing VIC 3029
AUSTRALIA

CONTENTS

PREFACE

This short book has several aims, particularly

➤ To give a concise introduction to the use of Argyris' *natural strain* concept in the Finite Element Method
➤ To give a concise introduction to the use of basis transformation in a 'natural' way on the sides of triangular elements to derive new elements.
➤ To give complete details of several new or improved elements applied to a wide range of problems, namely those of plane stress, thin plates, thick plates, flat and curved shell elements and elements for potential and viscous flow.

 The book contains much new work and material new to the literature, for example:
➤ Improved results are obtained for a quadratic basis thin plate element, now called the QBTP element, and it is further improved using small penalty factors.
➤ The new DFT2 element, which includes the drilling freedom, is a considerable improvement over its predecessor.
➤ The QBTP and DFT2 elements are combined to obtain a very useful flat or facet shell element.
➤ The element geometry of a 30 df doubly curved shell element is calculated using a natural approach, in turn allowing a completely natural formulation of the element with natural strains used for the flexural, shear and membrane strains and the radial and circumferential components of these.
➤ An accurate cubic element for potential flow, along with simple infinity conditions to model infinite domains similar to those developed for plane stress in 1978.
➤ An elegant element for viscous flow is obtained using selective integration by parts.
➤ The correct Cartesian and curvilinear 'large curvature corrections (LCCs) are derived for the first time.
➤ BASIC programs are given for the facet shell element, the doubly curved shell element, the viscous flow element, the curvilinear LCC and for optimization of viscous fluid flows.

As this list suggests, the work has a long history.

For example the coordinate transformation for the facet shell element of Chapter Six I developed during my PhD work in Cambridge in 1975-76.

In the next few years good deal of study of Argyris' work followed, resulting in the thick plate element of Chapter 7.

My first work on my 'large curvature correction' was in 1979.

The 'nested interpolation' method, i.e., application of basis transformation along the sides of triangular elements to obtain more convenient 'local' interpolations, was developed in my early months in Auckland in 1980, resulting in the QBTP and DFT1 elements. In the same period the first version of the doubly curved shell element of Chapter 8 was produced.

The improved QBTP element, including small penalty factors, was obtained in 1997.

The DFT2 element was obtained in 1998 and the resulting QBTP + DFT2 facet shell element was produced in 1999.

The cubic potential flow element and infinity conditions were produced circa 1998 and, further applications of this work to distribution and traffic flow networks were developed in 1997 and 2001-2 respectively.

In those years much other new work, for example development of the Patch Method, a hybrid FDM/FEM method, new analytical and FEM solutions for optimal triangular plates and application of FEM techniques in economics was done as well.

During writing of the first edition of this book in 2003 the simple 'chaining' of basis transformation for line elements in Section 2.5 was obtained, the somewhat tentative results of Chapter 8 for a spherical shell were obtained, and the formulation of the element for viscous flow in Equation 10.7 was simplified.

The author's early research was done in FORTRAN but a change to BASIC began in 1984, resulting in the short book *A Microcomputer Introduction to the Finite Element Method* (Pitman, Melbourne 1986; Heinemann, London, 1987). In 1988 I began using MegaBasic, and my still available 1992 OUP book *Finite Elements for Solids, Fluids, and Optimization* includes several MegaBasic programs.

In the present book VB version 5 is used (this fully compatible with VB6/7) because earlier BASICs are too slow for the shell element programs given in the book, and was also needed to obtain the 'large curvature correction' results of Table 8 in Appendix A.

Finally, I hope this short book and its examples of *naturals* and *basis transformation* prove of wide and continuing interest, and that some of the techniques, elements and programs in it will prove useful to readers.

Geoff Mohr, 2019

Chapter 1

INTRODUCTION

"I have no great intelligence, I have imagination",
"You are the hope of the future", John Argyris,
(said to the author by phone from Stuttgart circa 1998 & 1999).

1.1. New features of the book

This short book, though having to devote some time to introductory material, is a groundbreaking one, that is:

➢ It gives a concise introduction to Argyris' *natural strains* and their use in formulating finite elements elegantly and accurately.

➢ *Basis transformation,* that is, transformation of the global element freedoms to an alternative local set of freedoms is demonstrated, formulation of the element being simpler in terms of the local freedoms. For simple line elements it is shown that such basis transformation can be applied in a *chain* fashion.

➢ It gives a concise introduction to Mohr's *method of nested interpolations,* a basis transformation technique in which, typically, Cartesian displacement freedoms are transformed to *natural* freedoms defined parallel to the sides of triangular elements. Then interpolating along the element sides global slope freedoms are transformed to more convenient *natural slope* freedoms.

➢ The nested interpolation method is applied to form two nine freedom thin plate elements, one also using natural strains and being exactly equivalent to the classical BCIZ element, the other being perhaps the most accurate solution for this important problem.

➢ The nested interpolation method is applied to obtain two nine freedom plane stress elements incorporating the drilling freedom at each vertex. The second of these is particularly simple and passes patch tests.

➢ Nine freedom thin plate and plane stress elements obtained using the nested interpolation method are combined to obtain a very useful facet shell element. A complete program is given for this element.

3

➢ Natural strains are used to formulate an 18 freedom thick plate element.

➢ The latter is combined with the classical linear strain triangle for plane stress to obtain a curved shell element the differential geometry of which is also defined using the *natural* approach. The element code is given for this element.

➢ Alternative and equivalent derivations are given for a nine freedom potential flow element, using *natural* velocity freedoms parallel to its sides and simple 'single number' *infinity conditions* are then used to model potential field problems with infinite domains.

➢ An elegant 18 freedom element for viscous flow is simplified using *selective integration by parts* of the governing D.Es and small penalty factors. A program is given for this element.

➢ Steepest descent is used to optimize a FEM flow model and an introductory program given for this.

1.2. A brief history of the finite element method

The *Finite Element Method (FEM)* is now very widely used. This method is based on *matrix structural analysis* of aircraft structures in the mid 1940s, in which the *elements* of the structure were modeled mathematically by small matrices. For simple spar or *line elements* this could be done exactly (Livesley, 1953), but to model patches of the aircraft 'skin' approximate *lumping* techniques were used to represent them as, for example, three line elements to represent a triangular patch.

In 1954 Turner (Boeing), Clough (UC Berkeley), Martin (U Washington), and Topf (Boeing) produced the first paper on *continuum* finite elements at a meeting in New York and it was published in 1956 (Turner et al., 1956).

Argyris (U Stuttgart and U London), the best known pioneer of matrix structural analysis, provided a bridge between the 1940s 'lumping' work and these continuum elements with his *natural strains* on the sides of simplex elements, i.e. triangles in 2D and tetrahedra in 3D (Argyris, 1960).

Since then FEM has gone on to be applied to a wide range of problems in engineering and science, the main field of application being in *Computational Mechanics*, particularly the mechanics of solid buildings and manufactured products, including cars and aircraft.

In that time much effort has been directed at developing better element 'recipes', for example for the analysis of curved shell structures and shock fronts in supersonic air flows. Much effort too has been directed at *optimizing* FEM models, and the author has devoted much of over three decades to both these areas.

1.3. A brief introduction to the finite element method

The simplest finite elements are naturally discrete line elements of skeletal structures for which the element matrices can be derived from some of the basic equations of structural mechanics, as demonstrated in Section 2.1.

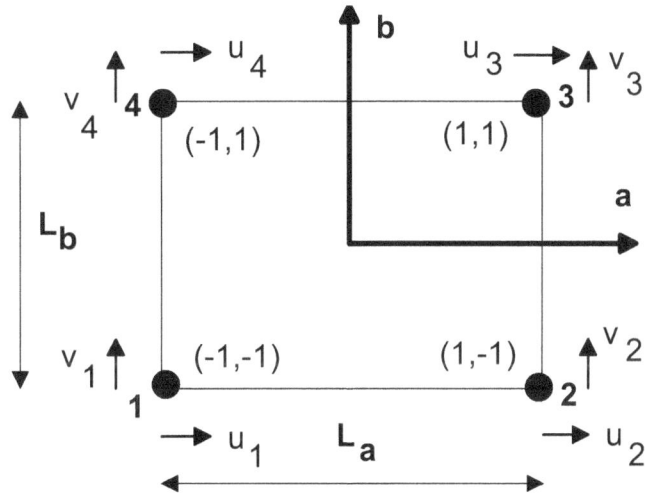

Figure 1.1. Bilinear element for plane stress

Figure 1.1 shows a more typical finite element, a rectangular element with translational freedoms u,v at each of four corner nodes. The nodal coordinates shown are local coordinates a,b parallel to the x,y axes, these being given by

$$a = 2x/L_a, \quad b = 2y/L_b \tag{1.1}$$

For a line element between nodes 1 and 2 the linear interpolation for the u displacement parallel to this side can be written by inspection as

$$u = \tfrac{1}{2}(1 - a)u_1 + \tfrac{1}{2}(1 + a)u_2 \tag{1.2}$$

5

Then the *blinear* interpolation for all four nodes is given by applying Eqn 1.2 in both the *a* and *b* directions and multiplying the results

$$u = \frac{1}{4}(1-a)(1-b)u_1 + \frac{1}{4}(1+a)(1-b)u_2 + \frac{1}{4}(1+a)(1+b)u_3 + \frac{1}{4}(1-a)(1-b)u_4$$

$$= \{f\}^t\{u\} = \sum f_i u_i = f_1 u_1 + f_2 u_2 + f_3 u_3 + f_4 u_4$$

$$(1.3)$$

where $\{u\}$ is a *vector* or column matrix of the nodal *u* values and $\{f\}^t$ is the transpose of the vector of *interpolation functions* f_i for each node, that is a row matrix of the f_i, and the standard rule of matrix multiplication yields the summation on the right hand side.

Applying the same interpolation to the *v* freedoms the direct strains at any point in the element can be expressed as

$$\varepsilon_x = \partial u/\partial x = [\{\partial f/\partial a\}^t(\partial a/\partial x) + \{\partial f/\partial b\}^t(\partial b/\partial x)]\{u\}^t$$

$$(1.4)$$

$$\varepsilon_y = \partial v/\partial y = [\{\partial f/\partial a\}^t(\partial a/\partial y) + \{\partial f/\partial b\}^t(\partial b/\partial y)]\{v\}^t \qquad (1.5)$$

and for the special case of a rectangular element with sides parallel to the axes $\partial a/\partial y = 0, \partial b/\partial x = 0$ and substituting the interpolation functions f_i of Equation 1.3 yields

$$\{\partial f/\partial a\}^t = \frac{1}{4}[-(1-b),\ (1-b),\ (1+b),\ -(1+b)]$$

$$(1.6)$$

$$\{\partial f/\partial b\}^t = \frac{1}{4}[-(1-a),\ -(1+a),\ (1+a),\ (1-a)]$$

$$(1.7)$$

These results are coded and deployed to form a 3×8 strain interpolation matrix *B* in the same manner as shown for the linear strain triangle in Equations 3.40 - 3.42:

$$\{\varepsilon\} = \{\partial u/x, \partial v\partial y, \partial v/\partial x + \partial u/\partial y\} = B\{d\} \qquad (1.8)$$

where $\{d\} = \{u_1, v_1, u_2, v_2, u_3, v_3, u_4, v_4\}$ is the vector of the element *freedoms* or nodal displacements.

Using *numerical* integration to evaluate Equations 1.6 and 1.7 and thence B at the four *Gauss points* (a,b) = $(\pm 1/\sqrt{3}, \pm 1/\sqrt{3})$, these being sufficient to integrate cubic terms exactly (Mohr, 1992), the strains are integrated over the element volume to give the total strain energy

$$U = \tfrac{1}{2}\{d\}^t[\int B^t D B \, dV]\{d\} = \tfrac{1}{2}\{d\}^t[\, \Sigma B^t D B \, (L_a L_b t / 4)]\{d\} = \tfrac{1}{2}\{d\}^t k \{d\}$$

(1.9)

where D is the modulus matrix (see Equation 3.49) which calculates the stresses from the strains and k is the *element stiffness matrix* (ESM).

Equating this internal work to the external work done by the loads $\{q\}$ applied at the nodes, that is $\{d\}^t\{q\}$, and minimizing with respect to $\{d\}$, the element equations are

$$\{q\} = k\{d\}$$

(1.10)

and summing this result for a collection of elements the *system equations* are

$$\boxed{\{Q\} = \Sigma\{q\} = [\Sigma k][\Sigma\{d\}] = K\{D\}}$$

(1.11)

where $\{Q\}$ are the loads on the system (at the nodes) and K is the *system stiffness matrix* (SSM), obtained by dividing the element matrices into sixteen 2×2 blocks and deploying these according the *global* node numbers of the element's nodes, this being a simple coding exercise.

If FEM can be summarized in one line Equation 1.11 does so and the author once used to write it at the top of the board at the beginning of each lecture to first time students of the subject.

Many examples of FEM follow in later chapters. Also as an introduction, Section 2.1 describes application of FEM to simple spar elements of a planar truss structure and Section 3.4 repeats the description above in more complete detail for the linear strain triangle.

In Chapter 9 simple elements for fluid and other flows are introduced. Here the flows are described by flow velocities which are calculated as derivatives of an interpolation of a potential function ϕ.

Here the problem is governed by a simple differential equation, which is reduced in order by integration by parts, yielding forcing terms at the element boundary in the process.

As an immediate example the equation governing steady state anisotropic heat flow is

$$\partial(\kappa_x \partial T/\partial x)/\partial x + \partial(\kappa_y \partial T/\partial y)\partial y + G = 0 \tag{1.12}$$

where $T = T(x,y)$ is the temperature distribution, κ_x and κ_y are the thermal conductivities in each axial direction and G is the heat generation per unit volume

Reducing the problem to one dimension with $G = 0$, substituting an interpolation for T, integrating over the element volume and using Galerkin weighting (with the interpolation functions) we obtain

$$A\kappa \int \{f\}\{f_{xx}\}^t \, dx \, \{T\} = \{0\} \quad \text{where } \{f_{xx}\} = \partial^2\{f\}/\partial x^2 \tag{1.13}$$

where A is the cross-sectional area of the element. Applying integration by parts to this result gives

$$A\kappa \int \{f_x\}\{f_x\}^t \, dx \, \{T\} = A\kappa \{f\} \, \partial T/\partial x \, | \tag{1.14}$$

where I denoted evaluation at the element boundary.

Then using a linear interpolation (of two nodal values) we have

$$\{f\} = \{1 - x/L, \ x/L\} \quad \text{and} \quad \{f_x\} = \{-1/L, \ 1/L\} \tag{1.15}$$

and substituting Equations 1.15 into Equation 1.14 we obtain

$$(A\kappa/L)\begin{bmatrix} 1 & -1 \\ -1 & 1 \end{bmatrix}\begin{Bmatrix} T_1 \\ T_2 \end{Bmatrix} = A\kappa \begin{Bmatrix} (\partial T/\partial x)_1 \\ (\partial T/\partial x)_2 \end{Bmatrix} \tag{1.16}$$

noting that the presence of the interpolation on the RHS of Equation 1.14 indicates simply that the nodal values of the variable interpolated are used for this boundary term. The resulting terms on the RHS of Equation 1.16 are the *inter element reactions* or fluxes and these sum to zero between elements except at the element boundary where they provide the forcing functions or 'loads' for the problem.

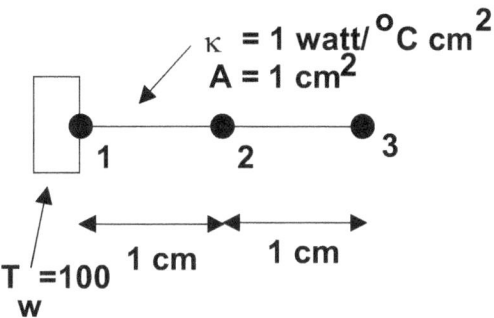

Figure 1.2. One dimensional heat flow problem

Figure 1.2 shows a simple 1-D heat flow example with only two elements. Then using Equation 1.16 the equations for the system are

$$\begin{bmatrix} 1 & -1 & 0 \\ -1 & 2 & -1 \\ 0 & -1 & 1 \end{bmatrix} \begin{Bmatrix} T_1 \\ T_2 \\ T_3 \end{Bmatrix} = \begin{Bmatrix} 0 \\ 0 \\ 0 \end{Bmatrix} \tag{1.17}$$

Imposing the boundary condition $T_1 = 100$ by multiplying the first column by this value and transposing it to the RHS, then removing the first row and column, we obtain the reduced problem

$$\begin{bmatrix} 2 & -1 \\ -1 & 2 \end{bmatrix} \begin{Bmatrix} T_2 \\ T_3 \end{Bmatrix} = \begin{Bmatrix} 100 \\ 0 \end{Bmatrix} \tag{1.18}$$

yielding the expected results $T_2 = T_3 = 100$ and, if we had put $T_3 = 0$ then the result would have been $T_2 = 50$, again the expected result.

The problem is, though almost trivial, a good example of FEM at its simplest though some of the mathematics involved when we begin with a differential equation will worry some readers at first.

Noting that the terms $\{f\}\{f\}^t$, which occur in many flow problems, are analogous to the energy product terms B^tB in solids problems, we have in this simple form a general method of dealing with most problems of mathematical modeling. The method remains fairly simple when we use the simplest possible elements but, of course, becomes a little more complicated when larger more complex elements are used.

1.4. References

Argyris JH, *Energy Theorems and Structural Analysis*, Butterworth, London 1960. Reprinted from Aircraft Engineering 1954-1955.

Livesley RK, Analysis of rigid frames by an electronic computer, *Engineering* 176 (1953) 230-238.

Mohr GA, *Finite Elements for Solids, Fluids, and Optimization*, Oxford University Press, Oxford, 1992.

Mohr GA, *Finite Elements & Optimization for Modern Management*, Amazon-Kindle, 2019.

Turner MJ, Clough RW, Martin HC, Topp LJ, Stiffness and deflection analysis of complex structures, *Jour. Aero. Sciences* 23 (1956) 805-823.

Chapter 2

LINE ELEMENTS AND BASIS TRANSFORMATION

2.1. Linear spar elements

Figure 2.1(a) shows a spar element with two *degrees of freedom,* horizontal and vertical displacement *u* and *v,* at a *node* at each end, thus having a total of *four freedoms.* Figure 2.1(b) shows a truss modeled by a number of these elements, subjected to a load at its end.

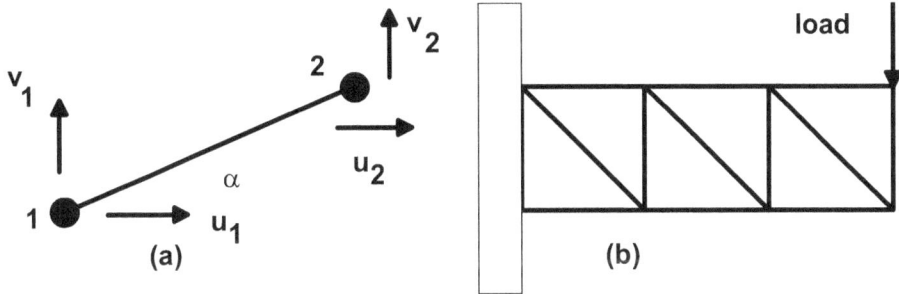

Figure 2.1. (a) Four freedom spar element; (b) Truss structure.

The four global element freedoms { *d* } shown can be transformed to local freedoms (*d**) parallel and perpendicular to the element using the transformation

$$\{d^*\} = \begin{Bmatrix} u_1{}^* \\ v_1{}^* \\ u_2{}^* \\ v_2{}^* \end{Bmatrix} = \begin{bmatrix} c & s & 0 & 0 \\ -s & c & 0 & 0 \\ 0 & 0 & c & s \\ 0 & 0 & -s & c \end{bmatrix} \begin{Bmatrix} u_1 \\ v_1 \\ u_2 \\ v_2 \end{Bmatrix} = T\{d\} \qquad (2.1)$$

where $c = \cos\alpha$, $s = \sin\alpha$ and α is the angle of inclination of the element from the horizontal *x* axis.

Applying one dimensional linear Lagrangian interpolation to the parallel or axial displacements at the ends of the element

$$u^* = \{f\}^t(u_i^*) = (1 - x^*/L)u_1^* + (x^*/L)u_2^* \qquad (2.2)$$

where $\{f\}$ is a vector of interpolation functions, here functions of the local coordinate x^* along the element length, and L is the element length.

The element extensional strain is given by

$$\varepsilon = d\{u^*\}/dx^* = B\{d^*\} = [-1/L,\ 0,\ 1/L,\ 0]\{d^*\} \qquad (2.3)$$

where B is the *strain interpolation matrix*.

Integrating over the element volume the strain energy of the element can thus be expressed as

$$U_e = (1/2)\int\{d^*\}^t B^t(E)\,B\{d^*\}\,d\bar{V} = (EAL/2)\{d^*\}^t B^t B\{d^*\} \qquad (2.4)$$

Equating this with the external work done by the nodal loads $\{q\}$ corresponding to the displacements $\{d\}$ and incorporating the transformation of Equation 2.1 we obtain

$$\{d\}^t\{q\} = (EAL/2)\{d\}^t T^t B^t B T\{d\} \qquad (2.5)$$

and minimizing with respect to the nodal displacements we obtain

$$\{q\} = \begin{Bmatrix} q_{x1} \\ q_{y1} \\ q_{x2} \\ q_{y2} \end{Bmatrix} = (EA/L)\begin{bmatrix} k_n & -k_n \\ -k_n & k_n \end{bmatrix}\begin{Bmatrix} u_1 \\ v_1 \\ u_2 \\ v_2 \end{Bmatrix} = k_e\{d\} \qquad (2.6)$$

$$\text{where } k_n = \begin{bmatrix} c^2 & -sc \\ -sc & s^2 \end{bmatrix} \text{ with } c = \cos\alpha,\ s = \sin\alpha \qquad (2.7)$$

k_e is the *element stiffness matrix (ESM)*, (EA/L) is the extensional stiffness of the element and $\{q\}$ and $\{d\}$ are its load and displacement *vectors*.

Finally the ESMs for each element are assembled into a *system stiffness matrix (SSM)* K to model the behaviour of the whole structure, symbolically stated as

$$\{Q\} = K\{D\} \text{ where } K = \Sigma k_e \qquad (2.8)$$

12

Specifying the loads on the structure $\{Q\}$ and the *boundary conditions,* that is the supporting points at which $u = 0$ and/or $v = 0$, the problem is solved by solving the *simultaneous equations* of Equation 2.8 to determine the nodal displacements $\{D\}$, then calculating the strains in each element of the structure from these using Equations 2.1 and 2.3.

2.2. Cubic beam elements

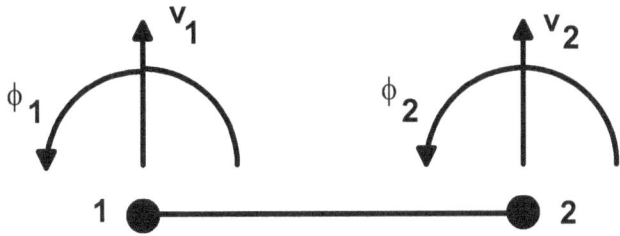

Figure 2.2. Four freedom beam element.

Figure 2.2 shows a four freedom beam element with transverse displacement and slope freedoms at each end. The cubic Hermitian interpolation for the element is obtained by writing the cubic polynomial

$$v = c_1 + c_2 s + c_3 s^2 + c_4 s^3 = \{c\}^t\{M\} \quad s = 0 \rightarrow 1 \qquad (2.9)$$

where $s = x/L$ is a *dimensionless coordinate* originating at node 1.

Differentiating Equation 2.9 with respect to s yields the interpolation for the slope as a *dimensionless derivative*

$$\phi^* = dv/ds = a_2 + 2a_3 s + 3a_4 s^2 \qquad (2.10)$$

and substituting the nodal values v_1, ϕ_1^*, v_2, ϕ_2^* on the left-hand sides of Equations 2.9 and 2.10 and the nodal coordinates $s = 0$ and $s = 1$ on the right-hand sides the four simultaneous equations thus obtained in matrix form are

$$\{d^*\} = \begin{Bmatrix} v_1 \\ \phi_1^* \\ v_2 \\ \phi_2^* \end{Bmatrix} = \begin{bmatrix} 1 & 0 & 0 & 0 \\ 0 & 1 & 0 & 0 \\ 1 & 1 & 1 & 1 \\ 0 & 1 & 2 & 3 \end{bmatrix} \begin{Bmatrix} c_1 \\ c_2 \\ c_3 \\ c_4 \end{Bmatrix} = C\{c\} \qquad (2.11)$$

If we require interpolation functions $\{f\}$ such that $v = \{f\}^t\{\ d^*\ \}$ than we have

$$v = \{f\}^t\{d^*\} = \{M\}^t\{c\} = \{M\}^t C^{-1}\{d^*\} \qquad (2.12/13)$$

so that the interpolation functions are given by

$$\{f\}^t = \{M\}^t C^{-1} = \left\{\begin{array}{c} 1 \\ s \\ s^2 \\ s^3 \end{array}\right\}^t \left[\begin{array}{cccc} 1 & 0 & 0 & 0 \\ 0 & 1 & 0 & 0 \\ -3 & -2 & 3 & -1 \\ 1 & 1 & -2 & 1 \end{array}\right] \qquad (2.14)$$

giving the interpolation functions as

$$\begin{aligned} f_1 &= 1 - 3s^2 + 2s^3 \ \text{(for } v_1) \\ f_2 &= s - 2s^2 + s^3 \ \text{(for } \phi_1{}^*) \\ f_3 &= 3s^2 - 2s^3 \ \text{(for } v_2) \\ f_4 &= s^3 - s^2 \ \text{(for } \phi_2{}^*) \end{aligned} \qquad (2.15)$$

noting that the global slope freedom $\phi = dv/dx = (dv/ds)(ds/dx) = \phi^*/L$.

Then the interpolation for the approximate (small displacement) curvature in the beam is given by

$$\begin{aligned} \chi &= d^2v/dx^2 = [d^2\{f\}^t/dx^2]\{d\} = [d^2\{f\}^t/ds^2](ds/dx)^2\{d\} \\ &= (1/L^2)[12s - 6, \ (6s - 4)L, \ 6 - 12s, (6s - 2)L]\{d\} = B\{d\} \end{aligned} \qquad (2.16)$$

and, using the same energy argument as for the spar element of the preceding section, the element stiffness matrix is given by

$$k_e = (EI/L^3) \ _0\!\int^1 B^t B\} \ dx = (EI/L^3)\left[\begin{array}{cccc} 12 & 6L & -12 & 6L \\ 6L & 4L^2 & -6L & 2L^2 \\ -12 & -6L & 12 & -6L \\ 6L & 2L^2 & -6L & 4L^2 \end{array}\right] \qquad (2.17)$$

This result can be combined with that of Equation 2.6 to yield the stiffness matrix for an inclined beam element and the result used to analyse structural frames (Mohr, 1992).

2.3. Basis transformation

A *linear space* V contains elements which obey the usual addition and multiplication laws of algebra, examples being:

1. The sets of real and complex numbers.

2. The set of infinite MacLaurin series.

3. A set of unit vectors.

4. The set of polynomials.

The elements of a bounded linear space obey the closure axioms

(a) For every pair x, y there is a unique element x + y.

(b) For every element x and real c there is a unique element cx.

A subset *S* of *V* which obeys the algebraic and closure axioms is called a *subspace* of V. Such a subspace may include linear combinations of the elements of *V* and subsets comprising alternative linear combinations of the same elements have the same *linear span*. For example the pairs of subsets

$$\{ i, j \}, \{ i, j, i + j, -j \}$$
$$\{1, x, x^2\}, \{1, 1+x, (1+x)^2\}$$

span the same subspace.

A finite set *S* is *independent* if all linear combinations $\Sigma c_i x_i = 0$ only if all $c_i = 0$. For example, the polynomial set $x_i = t^i, i = 0 \to n$ is obviously independent as the sum $\Sigma c_i x_i$ cannot vanish for $t \neq 0$ unless all the coefficients are zero.

A finite set *S* in a linear space *V* is called a *finite basis* for *V* if it is independent and spans *V*. If such a basis contains *n* elements then $n = \dim(V)$ is the dimension of *S*.

For the cubic beam element of Section 2.2, for example, we have a finite cubic basis $\{ M \} = \{ 1, x, x^2, x^3 \}$ which is clearly linearly independent with $\dim(\{ M \}) = 4$. Then when we obtain interpolation functions from the original polynomial this is, in fact, an exercise in basis transformation.

The coordinate transformation

$$\{x^*\} = \begin{Bmatrix} x^* \\ y^* \end{Bmatrix} = \begin{bmatrix} c & s \\ -s & c \end{bmatrix} \begin{Bmatrix} x \\ y \end{Bmatrix} = T\{x\} \tag{2.18}$$

is also a basis transformation.

If we express the original basis in linear combinatorial form, that is, each element of the basis is expressed as a vector with components x and y, we have

$$e = \{e_1, e_2\} = \{1x + 0y, 0x + 0y\} \qquad (2.19)$$

and the *inner product,,* a generalization of the vector dot product used to deal with general linear spaces, is used to define orthogonal spaces using the criteria

$$(e_i, e_j) = 0 \quad i \neq j \qquad (2.20)$$

$$(e_i, e_j) = 1 \quad i = j \qquad (2.21)$$

A basis possessing an inner product is a *Euclidean space* and it is *orthogonal* if it obeys Equation 2.20 and *orthonormal* if it obeys Equation 2.21 because its *Euclidean norm* $(e_i, e_i)^{1/2}$ is unity.

Then the transformation matrix T of Equation 2.18 is orthonormal because $T^{-1} = T^t$ and thus $T^{-1}T^t = I$, the identity matrix. Therefore the transformed basis $\{x^*, y^*\}$ is also orthonormal.

In the remainder of this chapter several other simple examples of basis transformation are given and in later chapters several very useful finite elements are obtained using basis transformation.

2.4. Basis transformation for a spar element

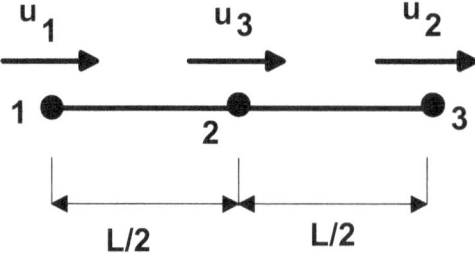

Figure 2.3. Quadratic spar element.

Considering a quadratic spar element with a node at its centre, as shown in Figure 2.3, the quadratic interpolation for its three freedoms is easily obtained by the inversion procedure used in Section 2.2 as

$$u = \{f\}^t\{u\} = \{1-3s+2s^2, 4s-4s^2, 2s^2-s\}\{u_1, u_3, u_2\} \quad s = 0 \to 1 \qquad (2.22)$$

and the extensional strain is given by

$$\epsilon = du/dx = (1/L)du/ds = (1/L)\{4s-3, \ 4-8s, \ 4s-1\}\{u\} \quad (2.23)$$

and putting $s = 0$ and $s = 1$ one obtains the strains at each end as

$$\begin{Bmatrix} \epsilon_1 \\ \epsilon_2 \end{Bmatrix} = (1/L)\begin{bmatrix} -3 & 4 & -1 \\ 1 & -4 & 3 \end{bmatrix}\{u\} = T_1\{u\} \quad (2.24)$$

Applying linear interpolation to these strains

$$\epsilon = \{f\}^t\{\epsilon\} = \{1-s, \ s\}\{\epsilon_1, \epsilon_2\} \quad (2.25)$$

and using this last result the kernel stiffness matrix for the 'local' linear element with strains at each end is

$$k^* = (EA)_0\int^L\{f\}\{f\}^t dx = (EAL)_0\int^1\{f\}\{f\}^t ds \quad (2.26)$$

$$= (EAL/6)\begin{bmatrix} 2 & 1 \\ 1 & 2 \end{bmatrix} \quad (2.27)$$

Then the global element stiffness matrix is given by congruent transformation with the basis transformation matrix of Equation 2.24

$$k = T_1^t \, k^* \, T_1 = (EA/6L)\begin{bmatrix} 14 & -16 & 2 \\ -16 & 32 & -16 \\ 2 & -16 & 14 \end{bmatrix} \quad (2.28)$$

which is the same result as that obtained by the usual method, that is using

$$k = (EA)_0\int^L \{df/dx\}\{df/dx\}^t dx \quad (2.29)$$

with $\{df/dx\}$ given by Equation 2.23.

The advantage of basis transformation here is that the kernel stiffness matrix is smaller and thence easier to formulate.

For two dimensional elements, however, 'complete' interpolations are not easy to obtain and later in the text basis transformation is used for two dimensional elements to transform the global freedoms to a local set more amenable to interpolation.

2.5. Basis transformation for a beam element

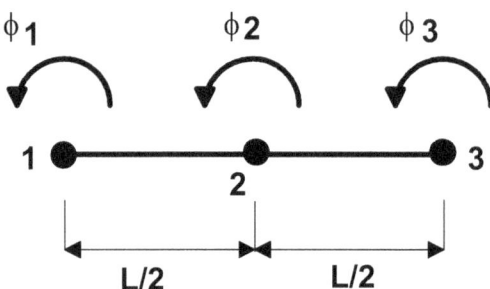

Figure 2.4. Quadratic 'local' element of that of Figure 2.2.

Figure 2.4 shows a quadratic beam element with three slope freedoms, obtained by applying *basis transformation* to the cubic element of Figure 2.2. This is done by differentiating Equations 2.15 to obtain the interpolation for the slope:

$$\phi = dv/dx = (1/L)dv/ds$$
$$= (1/L)\{6s^2 - 6s, \; L(3s^2 - 4s + 1), \; 6s - 6s^2. \; L(3s^2 - 2s)\}\{d\} \quad s = 0 \to 1 \tag{2.30}$$

Substituting s = 0, s = 1/2 and s = 1 in Equation 2.30 the required basis transformation is obtained as

$$\begin{Bmatrix} \phi_1 \\ \phi_2 \\ \phi_3 \end{Bmatrix} = \begin{bmatrix} 0 & 1 & 0 & 0 \\ -3/2L & -1/4 & 3/2L & -1/4 \\ 0 & 0 & 0 & 1 \end{bmatrix} \begin{Bmatrix} w_1 \\ \phi_1 \\ w_2 \\ \phi_2 \end{Bmatrix} = T_2\{d\} \tag{2.31}$$

The quadratic interpolation for the three 'local' slope freedoms of Figure 2.4 is that of Equation 2.22, that is

$$\phi = \{f\}'\{\phi\} = \{1 - 3s + 2s^2, \; 4s - 4s^2, \; 2s^2 - s\}\{\phi\} \tag{2.32}$$

giving the interpolation for the approximate curvature in the beam as

$$\chi = d\phi/dx = (1/L)d\phi/ds = (1/L)\{4s - 3, \; 4 - 8s, \; 4s - 1\}\{\phi\} \tag{2.33}$$

so that the stiffness matrix for the quadratic 'local' element is given by

$$k^* = EI \int_0^1 \{df/dx\}\{df/dx\}^t dx = (EI/L) \int_0^1 \{df/ds\}(df/ds)^t ds$$

$$= (EI/6L)\begin{bmatrix} 14 & -16 & 2 \\ -16 & 32 & -16 \\ 2 & -16 & 14 \end{bmatrix} \qquad (2.34)$$

and using congruent transformation with the basis transformation *matrix* T_2 of Equation 2.31 the global stiffness matrix is obtained as

$$k = T_2^t k^* T_2 \qquad (2.35)$$

yielding the same result as Equation 2.17.

Alternatively we can transform to a local two freedom element with a curvature freedom at each end. *Putting s = 0 and s = 1* in Equation 2.16 these curvatures are obtained as

$$\begin{Bmatrix} \chi_1 \\ \chi_2 \end{Bmatrix} = (1/L^2)\begin{bmatrix} -6 & -4L & 6 & -2L \\ 6 & 2L & -6 & 4L \end{bmatrix}\{d\} = T_3\{d\} \qquad (2.36)$$

and applying the linear interpolation

$$\chi = \{f\}^t\{\chi\} = \{(1-s),\ s\}^t\{\chi_1,\chi_2\} \qquad (2.37)$$

the kernel stiffness matrix is given by

$$k^* = (EI)\int_0^L \{f\}\{f\}^t dx = (EIL)\int_0^1 \{f\}\{f\}^t ds \qquad (2.38)$$

$$= (EIL/6)\begin{bmatrix} 2 & 1 \\ 1 & 2 \end{bmatrix} \qquad (2.39)$$

and the global element stiffness matrix is given by the congruent transformation

$$k = T_3^t k^* T_3 \qquad (2.40)$$

yielding the same result as Equation 2.17.

Indeed it is possible to achieve the latter basis transformation in two steps or *chain* it. This is done by using matrix T_2 of Equation 2.31 to transform from four displacement and slope freedoms to three slope freedoms, and then matrix T_1 of Equation 2.24 to transform to two curvature freedoms. To obtain the correct result (Equation 2.17) we then require

$$T_3 = T_1 T_2 \tag{2.41}$$

and indeed this is the case.

As a final example consider the local element of Figure 2.4 to obtain the geometric stiffness matrix for the four freedom beam element. Then using the interpolation of Equation 2.32, remembering that this applies to the slope of the beam, the kernel geometric stiffness matrix is obtained as

$$k_G{}^* = F \int_0^L \{f\}\{f\}^t dx = (FL) \int_0^1 \{f\}\{f\}^t ds = (FL/30) \begin{bmatrix} 4 & 1 & -1 \\ 2 & 16 & 2 \\ -1 & 2 & 4 \end{bmatrix}$$

$$\tag{2.42}$$

where F is the axial tensile force in the beam. Using the transformation matrix T_2 of Equation 2.31 the global geometric stiffness matrix is obtained as

$$k_G = T_2^t \, k_G{}^* \, T_2 = (F/30L) \begin{bmatrix} 36 & 3L & -36 & 3L \\ 3L & 4L^2 & -3L & -L^2 \\ -36 & -3L & 36 & -3L \\ 3L & -L^2 & -3L & 4L^2 \end{bmatrix} \tag{2.43}$$

which is the same result as obtained by the conventional derivation, that is

$$k_G = F \int_0^L \{df/dx\}\{df/dx\}^t dx \tag{2.44}$$

now using the differentiated cubic interpolation of Equation 2.30.

2.6. Conclusion

As the numerous examples given here show, basis transformation is an important technique which is very widely applicable in the finite element method.

In following chapters interpolations are applied to the sides of triangular elements for which satisfactory 'direct' interpolations do not exist to obtain local element freedoms for which standard interpolations can be used.

2.7. References

[1] Mohr GA, Finite element formulation by nested interpolations: application to a quadrilateral thin plate bending element, *Trans. IEAust CE25* (1983) 211-218.

[2] Mohr GA, *Finite Elements for Solids, Fluids, and Optimization*, Oxford University Press, Oxford, 1992.

2. Line Elements and Basis Transformation

Chapter 3

NATURAL COORDINATES AND STRAINS

3.1. Areal coordinates for triangles

We have already encountered Lagrangian and Hermitian interpolation in line elements in Chapter 2 and, as shown in Section 1.3, interpolation in rectangles and quadrilaterals can be accomplished by combination of one dimensional interpolations.

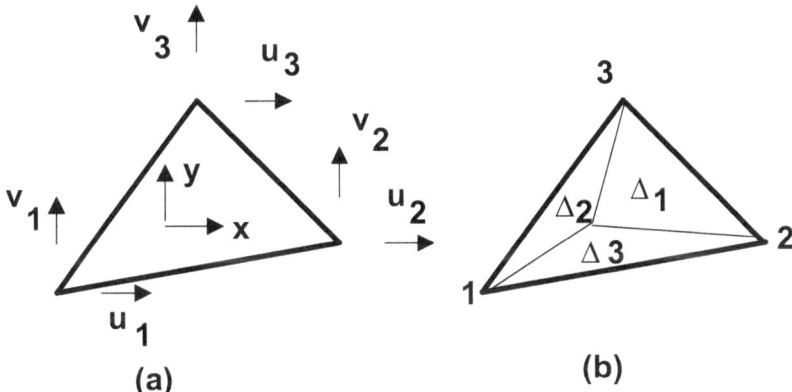

Figure 3.1. (a) Constant strain triangle. (b) Areal coordinates.

The focus of the present work is triangular elements and for these *natural coordinates* can be derived by writing the linear interpolation for the classical constant strain triangle (CST) element

$$u = \{c\}^t\{M\} = c_1 + c_2 x + c_3 y \tag{3.1}$$

and substituting the coordinates at the vertices of a triangle and writing the results in matrix form gives

$$\begin{Bmatrix} u_1 \\ u_2 \\ u_3 \end{Bmatrix} = \begin{bmatrix} 1 & x_1 & y_1 \\ 1 & x_2 & y_2 \\ 1 & x_3 & y_3 \end{bmatrix} \begin{Bmatrix} c_1 \\ c_2 \\ c_3 \end{Bmatrix} = C\{c\} \tag{3.2}$$

Then writing the interpolation in terms of interpolation functions

$$u = \{f\}^t \{u\} \quad \text{where} \quad \{f\} = (C^{-1})^t \{M\} \tag{3.3}$$

where

$$C^{-1} = (1/|C|) \begin{bmatrix} a_1 & a_2 & a_3 \\ -y_{32} & -y_{13} & -y_{21} \\ x_{32} & x_{13} & x_{21} \end{bmatrix} \tag{3.4}$$

in which

$$x_{32} = x_3 - x_2, \; y_{32} = y_3 - y_2 \quad \text{etc.}$$
$$a_1 = x_2 y_3 - x_3 y_2 = 2\Delta/3, \quad a_2 = x_3 y_1 - x_1 y_3 = 2\Delta/3, \quad a_3 = x_1 y_2 - x_2 y_1 = 2\Delta/3$$
$$|C| = a_1 + a_2 + a_3 = 2\Delta \tag{3.5}$$

giving the interpolation functions as

$$f_1 = L_1 = (a_1 - y_{32}x + x_{32}y)/2\Delta$$
$$f_2 = L_2 = (a_2 - y_{13}x + x_{13}y)/2\Delta \tag{3.6}$$
$$f_3 = L_3 = (a_3 - y_{21}x + x_{21}y)/2\Delta$$

and the element area Δ is usually calculated from the formula

$$2\Delta = |x_{21}y_{32} - x_{32}y_{21}| \tag{3.7}$$

or one of its two permutations.

Then L_1, L_2, L_3 in Equations 3.6 are the *areal coordinates* and these are the ratios of the areas shown in Figure 3.1(b) to the total area, that is

$$L_1 = \Delta_1/\Delta, \; L_2 = \Delta_2/\Delta, \; L_3 = \Delta_3/\Delta \tag{3.8}$$

from which follows the identity

$$L_1 + L_2 + L_3 = 1 \tag{3.9}$$

3.2. Quadratic and cubic Lagrangian triangles

The classical 6 freedom constant strain triangle is very easy to formulate, and even more so using areal coordinates (Mohr, 1992). The twelve freedom linear strain triangle (LST) shown in Figure 3.2 is much more accurate and, indeed, elements of at least linear strain accuracy are generally recommended for practical finite element analysis.

As shown, the LST has additional nodes at the middle of each side for which the areal coordinates take the simple values shown, greatly simplifying derivation of the interpolation functions.

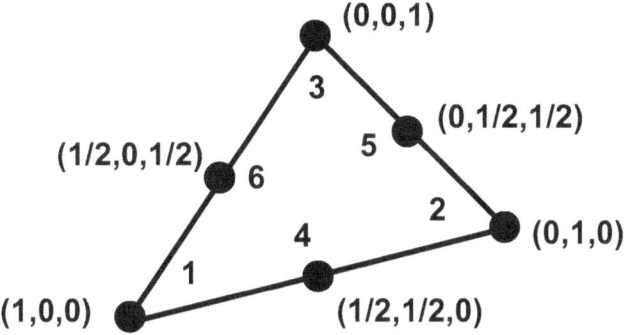

Diplacement freedoms are u, v at each node

Figure 3.2. Linear strain triangle, showing nodal area coordinates.

Writing the interpolation as a quadratic polynomial in the areal coordinates

$$u = c_1 L_1^2 + c_2 L_2^2 + c_3 L_3^2 + c_4 L_1 L_2 + c_5 L_2 L_3 + c_6 L_3 L_1 = \{M\}^t \{c\} \tag{3.10}$$

and substituting the nodal u values on the left side and the nodal coordinates on the right to form an interpolation matrix C

$$\begin{Bmatrix} u_1 \\ u_2 \\ u_3 \\ u_4 \\ u_5 \\ u_6 \end{Bmatrix} = \begin{bmatrix} 1 & 0 & 0 & 0 & 0 & 0 \\ 0 & 1 & 0 & 0 & 0 & 0 \\ 0 & 0 & 1 & 0 & 0 & 0 \\ 1/4 & 1/4 & 0 & 1/4 & 0 & 0 \\ 0 & 1/4 & 1/4 & 0 & 1/4 & 0 \\ 1/4 & 0 & 1/4 & 0 & 0 & 1/4 \end{bmatrix} \begin{Bmatrix} c_1 \\ c_2 \\ c_3 \\ c_4 \\ c_5 \\ c_6 \end{Bmatrix} = C\{c\} \tag{3.11}$$

and inverting matrix C the interpolation functions are given by

$$\{f\}^t = \{M\}^t C^{-1} = \{M\}^t \begin{bmatrix} 1 & 0 & 0 & 0 & 0 & 0 \\ 0 & 1 & 0 & 0 & 0 & 0 \\ 0 & 0 & 1 & 0 & 0 & 0 \\ -1 & -1 & 0 & 4 & 0 & 0 \\ 0 & -1 & -1 & 0 & 4 & 0 \\ -1 & 0 & -1 & 0 & 0 & 4 \end{bmatrix} \tag{3.12}$$

so that, using the identity of Equation 3.9 to simplify the first three functions, we obtain

$$f_1 = L_1^2 - L_1 L_2 - L_3 L_1 = L_1^2 - L_1(L_2 + L_3) = L_1^2 - L_1(1 - L_1) = 2L_1^2 - 1$$

$$f_2 = 2L_2^2 - 1$$

$$f_3 = 2L_3^2 - 1$$

$$f_4 = 4L_1 L_2, \quad f_5 = 4L_2 L_3, \quad f_6 = 4L_3 L_1$$

$$\tag{3.13}$$

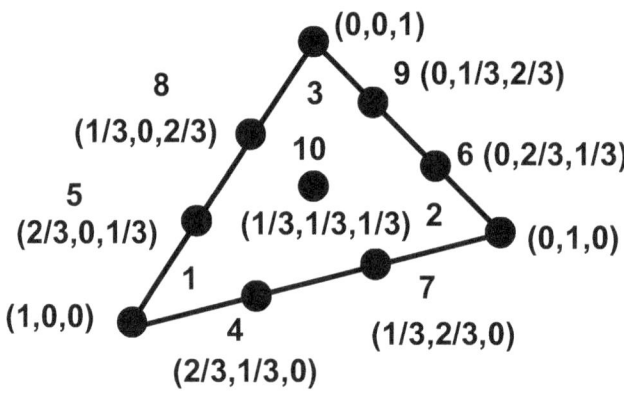

Figure 3.3. Cubic triangle.

Figure 3.3 shows the cubic Lagrangian triangle element for which the modes of the areal coordinate polynomial interpolation are

$$\{M\} = \{L_1^3, \ L_2^3, \ L_3^3, \ L_1^2 L_2, L_1^2 L_3, \ L_2^2 L_3, \ L_2^2 L_1, \ L_3^2 L_1, \ L_3^2 L_2, \ L_1 L_2 L_3\} \tag{3.14}$$

Substituting the nodal coordinates shown in Figure. 3.3 the interpolation *matrix* C is obtained as

$$C = (1/27) \begin{bmatrix} 27 & 0 & 0 & 0 & 0 & 0 & 0 & 0 & 0 & 0 \\ 0 & 27 & 0 & 0 & 0 & 0 & 0 & 0 & 0 & 0 \\ 0 & 0 & 27 & 0 & 0 & 0 & 0 & 0 & 0 & 0 \\ 8 & 1 & 0 & 4 & 0 & 0 & 2 & 0 & 0 & 0 \\ 8 & 0 & 1 & 0 & 4 & 0 & 0 & 2 & 0 & 0 \\ 0 & 8 & 1 & 0 & 0 & 4 & 0 & 0 & 2 & 0 \\ 1 & 8 & 0 & 2 & 0 & 0 & 4 & 0 & 0 & 0 \\ 1 & 0 & 8 & 0 & 2 & 0 & 0 & 4 & 0 & 0 \\ 0 & 1 & 8 & 0 & 0 & 2 & 0 & 0 & 4 & 0 \\ 1 & 1 & 1 & 1 & 1 & 1 & 1 & 1 & 1 & 1 \end{bmatrix} \tag{3.15}$$

Inverting C the interpolation functions are given by

$$\{f\}^t = \{M\}^t C^{-1} = \{M\}^t(1/2) \begin{bmatrix} 2 & 0 & 0 & 0 & 0 & 0 & 0 & 0 & 0 & 0 \\ 0 & 2 & 0 & 0 & 0 & 0 & 0 & 0 & 0 & 0 \\ 0 & 0 & 2 & 0 & 0 & 0 & 0 & 0 & 0 & 0 \\ -5 & 2 & 0 & 18 & 0 & 0 & -9 & 0 & 0 & 0 \\ -5 & 0 & 2 & 0 & 18 & 0 & 0 & -9 & 0 & 0 \\ 0 & -5 & 2 & 0 & 0 & 18 & 0 & 0 & -9 & 0 \\ 2 & -5 & 0 & -9 & 0 & 0 & 18 & 0 & 0 & 0 \\ 2 & 0 & -5 & 0 & -9 & 0 & 0 & 18 & 0 & 0 \\ 0 & 2 & -5 & 0 & 0 & -9 & 0 & 0 & 18 & 0 \\ 4 & 4 & 4 & -9 & -9 & -9 & -9 & -9 & -9 & 54 \end{bmatrix}$$

$$\tag{3.16}$$

giving, after simplification using Equation 3.9, the interpolation functions

$$f_1 = 4.5L_1^3 - 4.5L_1^2 + L_1, \quad f_2 = 4.5L_2^3 - 4.5L_2^2 + L_2, \quad f_3 = 4.5L_3^3 - 4.5L_3^2 + L_3$$
$$f_4 = 13.5L_1^2 L_2 - 4.5L_1 L_2, \quad f_5 = 13.5L_1^2 L_3 - 4.5L_1 L_3$$
$$f_6 = 13.5L_2^2 L_3 - 4.5L_2 L_3, \quad f_7 = 13.5L_2^2 L_1 - 4.5L_2 L_1$$
$$f_8 = 13.5L_3^2 L_1 - 4.5L_3 L_1, \quad f_9 = 13.5L_3^2 L_2 - 4.5L_3 L_2$$
$$f_{10} = 27L_1 L_2 L_3$$

$$\tag{3.17}$$

Note that in Equation 3.10 the modes for the interpolation have been written in the homogeneous form

$$L_1^a L_2^b L_3^c \quad where \ a + b + c = 2 \tag{3.18}$$

but that one can equally well follow the practice for Cartesian polynomials and use an ascending polynomial, that is for the quadratic triangle using

$$\{M\} = \{L_1, \ L_2, \ L_3, \ L_1 L_2, \ L_2 L_3, \ L_3 L_1\} \tag{3.19}$$

and this approach leads to exactly the same interpolation functions for both the quadratic and cubic Lagrangian elements.

3.3. Hermitian triangular element

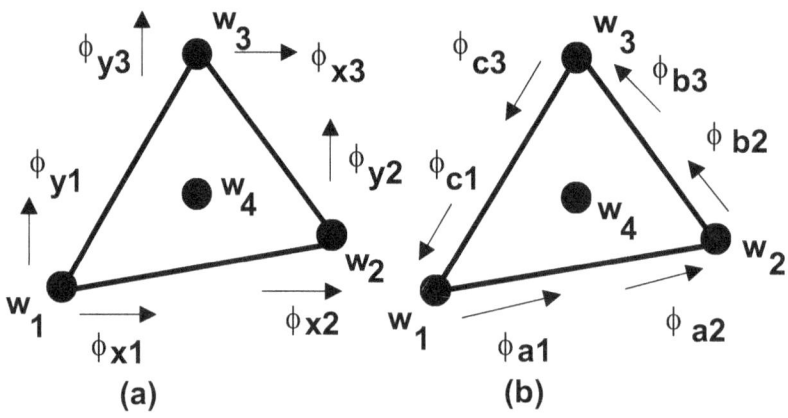

Figure 3.4. Hermitian triangle: (a) global freedoms, (b) local freedoms.

Figure 3.4(a) shows a ten freedom cubic Hermitian triangle. Derivation of the interpolation functions for this is much simplified if the Cartesian slope freedoms are replaced by *natural slopes* parallel to the element sides, as shown in Figure 3.4(b).

The natural slopes are defined as *dimensionless derivatives* with respect to coordinates *a, b* and *c* measured along each side of the element. Thus parallel to side 12 we define the natural slope

$$\phi_a = L_a(\partial w / \partial a) = L_a[(\partial w / \partial x)(\partial x / \partial a) + (\partial w / \partial y)(\partial y / \partial a)]$$
$$= L_a(c_{ax}\phi_x + c_{ay}\phi_y) = x_{21}\phi_x + y_{21}\phi_y \tag{3.20}$$

where c_{ax}, c_{ay} are the direction cosines of side 12 and are given by

$$c_{ax} = (x_2 - x_1)/L_a = x_{21}/L_a, \quad c_{ay} = (y_2 - y_1)/L_a = y_{21}/L_a \qquad (3.21)$$

Similarly

$$\phi_b = x_{32}\phi_x + y_{32}\phi_y, \quad \phi_c = x_{13}\phi_x + y_{13}\phi_y \qquad (3.22)$$

and thus the transformation from the global freedoms of Figure 3.4(a) to the local freedoms of Figure 3.4(b) is

$$\{d_N\} = \begin{Bmatrix} w_1 \\ w_2 \\ w_3 \\ \phi_{a1} \\ \phi_{c1} \\ \phi_{b2} \\ \phi_{a2} \\ \phi_{c3} \\ \phi_{b3} \\ w_4 \end{Bmatrix} = \begin{bmatrix} 1 & 0 & 0 & 0 & 0 & 0 & 0 & 0 & 0 & 0 \\ 0 & 0 & 0 & 1 & 0 & 0 & 0 & 0 & 0 & 0 \\ 0 & 0 & 0 & 0 & 0 & 0 & 1 & 0 & 0 & 0 \\ 0 & x_{21} & y_{21} & 0 & 0 & 0 & 0 & 0 & 0 & 0 \\ 0 & x_{13} & y_{13} & 0 & 0 & 0 & 0 & 0 & 0 & 0 \\ 0 & 0 & 0 & 0 & x_{32} & y_{32} & 0 & 0 & 0 & 0 \\ 0 & 0 & 0 & 0 & x_{21} & y_{21} & 0 & 0 & 0 & 0 \\ 0 & 0 & 0 & 0 & 0 & 0 & x_{13} & y_{13} & 0 \\ 0 & 0 & 0 & 0 & 0 & 0 & x_{32} & y_{32} & 0 \\ 0 & 0 & 0 & 0 & 0 & 0 & 0 & 0 & 1 \end{bmatrix} \begin{Bmatrix} w_1 \\ \phi_{x1} \\ \phi_{y1} \\ w_2 \\ \phi_{x2} \\ \phi_{y2} \\ w_3 \\ \phi_{x3} \\ \phi_{y3} \\ w_4 \end{Bmatrix} = T\{d\}$$

$$(3.23)$$

and a local kernel element matrix is determined in terms of the local *natural* freedoms and then congruently transformed using matrix T to obtain the final global element matrix.

Writing the displacement interpolation with the first three modes in 'ascending' form [because in the usual applications to thin plate bending these first three modes are rigid body modes and can be omitted from calculations]

$$w = \{c\}^t M\} = \{L_1, L_2, L_3, L_1^2 L_2, L_1^2 L_3, L_2^2 L_3, L_2^2 L_1, L_3^2 L_1, L_3^2 L_2, L_1 L_2 L_3\}$$
$$(3.24)$$

the natural slopes are expressed as dimensionless derivatives of this

$$\phi_a = L_{21}(\partial w/\partial a) = L_{21}[(\partial w/\partial L_1)(\partial L_1/\partial a) + (\partial w/\partial L_2)(\partial L_2/\partial a)]$$

$$= \partial w/\partial L_2 - \partial w/\partial L_1$$

$$(3.25)$$

and using cyclic progression

$$\phi_b = \partial w/\partial L_3 - \partial w/\partial L_2, \quad \phi_c = \partial w/\partial L_1 - \partial w/\partial L_3 \tag{3.26}$$

Applying these results to the cubic modes of Equation 3.24 and substituting the nodal area coordinates gives the interpolation matrix

$$\{d_N\} = \begin{Bmatrix} w_1 \\ w_2 \\ w_3 \\ \phi_{a1} \\ \phi_{c1} \\ \phi_{b2} \\ \phi_{a2} \\ \phi_{c3} \\ \phi_{b3} \\ w_4 \end{Bmatrix} = \begin{bmatrix} 1 & 0 & 0 & 0 & 0 & 0 & 0 & 0 & 0 & 0 \\ 0 & 1 & 0 & 0 & 0 & 0 & 0 & 0 & 0 & 0 \\ 0 & 0 & 1 & 0 & 0 & 0 & 0 & 0 & 0 & 0 \\ -1 & 1 & 0 & 1 & 0 & 0 & 0 & 0 & 0 & 0 \\ 1 & 0 & -1 & 0 & -1 & 0 & 0 & 0 & 0 & 0 \\ 0 & -1 & 1 & 0 & 0 & 1 & 0 & 0 & 0 & 0 \\ -1 & 1 & 0 & 0 & 0 & 0 & -1 & 0 & 0 & 0 \\ 1 & 0 & -1 & 0 & 0 & 0 & 0 & 1 & 0 & 0 \\ 0 & -1 & 1 & 0 & 0 & 0 & 0 & 0 & -1 & 0 \\ \frac{1}{3} & \frac{1}{3} & \frac{1}{3} & \frac{1}{27} & \frac{1}{27} & \frac{1}{27} & \frac{1}{27} & \frac{1}{27} & \frac{1}{27} & \frac{1}{27} \end{bmatrix} \{c\} = C\{c\}$$

$$\tag{3.27}$$

giving on inversion

$$\{f\}^t = \begin{Bmatrix} L_1^3 \\ L_2^3 \\ L_3^3 \\ L_1^2 L_2 \\ L_1^2 L_3 \\ L_2^2 L_3 \\ L_2^2 L_1 \\ L_3^2 L_1 \\ L_3^2 L_2 \\ L_1 L_2 L_3 \end{Bmatrix}^t \begin{bmatrix} 1 & 0 & 0 & 0 & 0 & 0 & 0 & 0 & 0 & 0 \\ 0 & 1 & 0 & 0 & 0 & 0 & 0 & 0 & 0 & 0 \\ 0 & 0 & 1 & 0 & 0 & 0 & 0 & 0 & 0 & 0 \\ 1 & -1 & 0 & 1 & 0 & 0 & 0 & 0 & 0 & 0 \\ 1 & 0 & -1 & 0 & -1 & 0 & 0 & 0 & 0 & 0 \\ 0 & 1 & -1 & 0 & 0 & 1 & 0 & 0 & 0 & 0 \\ -1 & 1 & 0 & 0 & 0 & 0 & -1 & 0 & 0 & 0 \\ -1 & 0 & 1 & 0 & 0 & 0 & 0 & 1 & 0 & 0 \\ 0 & -1 & 1 & 0 & 0 & 0 & 0 & 0 & -1 & 0 \\ -9 & -9 & -9 & -1 & 1 & -1 & 1 & -1 & 1 & 27 \end{bmatrix} = \{M\}^t C^{-1}$$

$$\tag{3.28}$$

so that the interpolation functions are

$$f_1 = L_1 + L_1^2 L_2 + L_1^2 L_3 - L_2^2 L_1 - L_3^2 L_1 - 9L_1 L_2 L_3$$
$$f_2 = L_2 + L_2^2 L_3 + L_2^2 L_1 - L_3^2 L_2 - L_1^2 L_2 - 9L_1 L_2 L_3$$
$$f_3 = L_3 + L_3^2 L_1 + L_3^2 L_2 - L_1^2 L_3 - L_2^2 L_3 - 9L_1 L_2 L_3$$
$$f_4 = L_1^2 L_2 - L_1 L_2 L_3, \quad f_5 = -L_1^2 L_3 + L_1 L_2 L_3$$
$$f_6 = L_2^2 L_3 - L_1 L_2 L_3, \quad f_7 = -L_2^2 L_1 + L_1 L_2 L_3$$
$$f_8 = L_3^2 L_1 - L_1 L_2 L_3, \quad f_9 = -L_3^2 L_2 + L_1 L_2 L_3$$
$$f_{10} = 27 L_1 L_2 L_3$$

$$(3.29)$$

3.4. The linear strain triangle

The quadratic element of Figure 3.2 was originally applied to plane stress problems and thus called the linear strain triangle (LST).

As the strains involve only first derivatives the Cartesian derivatives can be calculated by applying the chain rule to the displacement interpolation

$$\varepsilon_x = \partial u / \partial x = (\partial u / \partial L_1)(\partial L_1 / \partial x) + (\partial u / \partial L_2)(\partial L_2 / \partial x) + (\partial u / \partial L_3)(\partial L_3 / \partial x)$$
$$(3.30)$$

and substituting the interpolation $u = \{f\}^t \{u\}$ we obtain

$$\varepsilon_x = [\{\partial f / \partial L_1\}^t (\partial L_1 / \partial x) + \{\partial f / \partial L_2\}^t (\partial L_2 / \partial x) + \{\partial f / \partial L_3\}^t (\partial L_3 / \partial x)] \{u\}$$

$$(3.31)$$

where $\quad \{\partial f / \partial L_1\} = \{\partial f_1 / \partial L_1, \partial f_2 / \partial L_1, \ldots \ldots \partial f_6 / \partial L_1\}$

$$(3.32)$$

The interpolation functions are Equations 3.13, that is

$$f_1 = 2L_1^2 - 1, f_2 = 2L_2^2 - 1, f_3 = 2L_3^2 - 1, f_4 = 4L_1 L_2, f_5 = 4L_2 L_3, f_6 = 4L_3 L_1$$
$$(3.33)$$

Then an interpolation matrix S for the required local derivatives is readily obtained by differentiating Equations 3.33

$$S = \begin{bmatrix} \{\partial f/\partial L_1\}^t \\ \{\partial f/\partial L_2\}^t \\ \{\partial f/\partial L_3\}^t \end{bmatrix} = \begin{bmatrix} 4L_1-1 & 0 & 0 & 4L_2 & 0 & 4L_3 \\ 0 & 4L_2-1 & 0 & 4L_1 & 4L_3 & 0 \\ 0 & 0 & 4L_3-1 & 0 & 4L_2 & 4L_1 \end{bmatrix} \qquad (3.34)$$

and, using the areal coordinate definitions of Equations 3.6, the local area coordinate derivatives are transformed to Cartesian derivatives

$$\begin{bmatrix} \{\partial f/\partial x\}^t \\ \{\partial f/\partial y\}^t \end{bmatrix} = \begin{bmatrix} \partial L_1/\partial x & \partial L_2/\partial x & \partial L_3/\partial x \\ \partial L_1/\partial y & \partial L_2/\partial y & \partial L_3/\partial y \end{bmatrix} S = (1/2\Delta)\begin{bmatrix} -y_{32} & -y_{13} & -y_{21} \\ x_{32} & x_{13} & y_{21} \end{bmatrix} S = G$$

$$(3.35)$$

For illustrative purposes denoting the resulting matrix G as

$$G = \begin{bmatrix} g_{11} & g_{12} & g_{13} & g_{14} & g_{15} & g_{16} \\ g_{21} & g_{22} & g_{23} & g_{24} & g_{25} & g_{26} \end{bmatrix} \qquad (3.36)$$

the final strain interpolation matrix is given as

$$\begin{Bmatrix} \partial u/\partial x \\ \partial v/\partial y \\ \partial u/\partial y + \partial v/\partial x \end{Bmatrix} = \begin{bmatrix} g_{11} & 0 & g_{12} & 0 & g_{13} & 0 & g_{14} & 0 & g_{15} & 0 & g_{16} & 0 \\ 0 & g_{21} & 0 & g_{22} & 0 & g_{23} & 0 & g_{24} & 0 & g_{25} & 0 & g_{26} \\ g_{21} & g_{11} & g_{22} & g_{12} & g_{23} & g_{13} & g_{24} & g_{14} & g_{25} & g_{15} & g_{26} & g_{16} \end{bmatrix} \{d\} = B\{d\}$$

$$(3.37)$$

The final stiffness matrix is given by using numerical integration at the midside nodes (with weights $= 1/3$) which exactly integrates quadratic terms, as required:

$$k = \Sigma\, B^t D B\,(\Delta t/3) \qquad (3.38)$$

where D is the modulus matrix

$$D = Et/(1 - v^2)\begin{bmatrix} 1 & v & 0 \\ 0 & 1 & 0 \\ 0 & 0 & (1-v)/2 \end{bmatrix}$$

in which E is Young's modulus, t is the element thickness and v is Poisson's ratio.

The element stresses are calculated using

$$\{\sigma\} = DB\{d\} \tag{3.39}$$

at the nodes and generally it is advisable to use average nodal stresses, that is average the stresses given by the elements impinging at a node.

3.5. Isoparametric formulation of the LST element

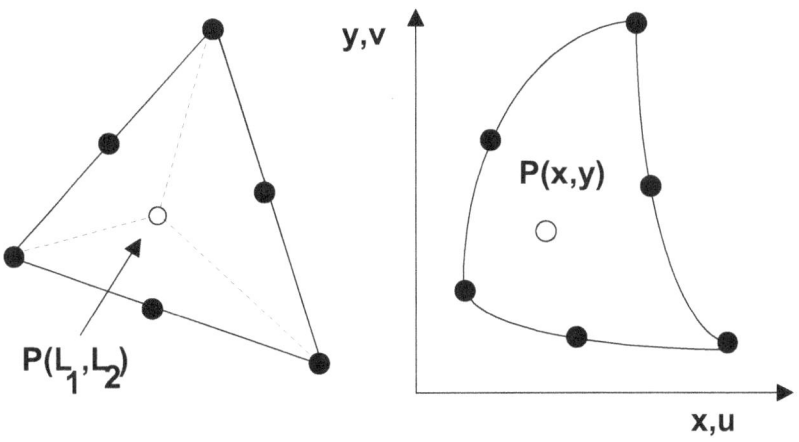

Figure 3.5. Isoparametric quadratic triangle element.

Figure 3.5 shows the LST element mapped from isoparametric form with curved sides in the x-y plane to straight sided form in areal coordinates.

Applying the interpolation functions of Equations 3.33 to both the displacement freedoms and coordinates

$$u = \Sigma f_i u_i, \; v = \Sigma f_i v_i, \; x = \Sigma f_i x_i, \; y = \Sigma f_i y_i, \quad i = 1 \rightarrow 6 \tag{3.40}$$

Eliminating L_3 from Eqns 3.33 using the identity $L_3 = 1 - L_1 - L_2$ and then differentiating with respect to L_1 and L_2 we obtain (Ergatoudis et al, 1968)

$$S = \begin{bmatrix} \partial\{f\}^t/\partial L_1 \\ \partial\{f\}^t/\partial L_2 \end{bmatrix}$$

$$= \begin{bmatrix} 4L_1 - 1 & 0 & 4L_1 + 4L_2 - 3 & 4L_2 & -4L_2 & 4 - 8L_1 - 4L_2 \\ 0 & 4L_2 - 1 & 4L_1 + 4L_2 - 3 & 4L_1 & 4 - 4L_1 - 8L_2 & -4L_1 \end{bmatrix}$$

$$\tag{3.41}$$

Then a Jacobian matrix is calculated numerically at each integration point

$$J = \begin{bmatrix} \partial x/\partial L_1 & \partial y/\partial L_1 \\ \partial x/\partial L_2 & \partial y/\partial L_2 \end{bmatrix} = S[\,\{x\}\ \{y\}\,] \tag{3.42}$$

giving the Cartesian derivatives as

$$\begin{Bmatrix} \partial u/\partial x \\ \partial u/\partial y \end{Bmatrix} = J^{-1}\,S\{u\} = G\{u\} \tag{3.43}$$

The elements of matrix G are deployed to form the strain-displacement matrix B in the manner shown in Equation 3.37 and the element stiffness matrix is given by three point integration at the midside nodes as

$$k = \Sigma\ B^t DB\,(t\ |\,J\,|_{abs}/6) \tag{3.44}$$

noting that $\Sigma\,|\,J\,|_{abs}$ gives twice the element area when area coordinates are used.

3.6. Natural strains

Natural strains in triangles are calculated parallel to the element sides. Then for plane stress problems the natural strains are calculated as the rate of change in displacement parallel to each side with respect to a natural coordinate along the side in the same way as natural slopes were calculated in the preceding section, that is

$$\varepsilon_a = d\delta_a/da = (\partial\delta_a/\partial L_1)(\partial L_1/\partial a) + (\partial\delta_a/\partial L_2)(\partial L_2/\partial a)$$
$$= [\partial\delta_a/\partial L_2 - \partial\delta_a/\partial L_1]/L_a \tag{3.45}$$
$$\varepsilon_b = d\delta_b/db = [\partial\delta_b/\partial L_3 - \partial\delta_b/\partial L_2]/L_b$$
$$\varepsilon_c = d\delta_c/dc = [\partial\delta_c/\partial L_1 - \partial\delta_c/\partial L_3]/L_c$$

First, therefore, the Cartesian nodal freedoms u, v must be transformed to natural values in the same way as for natural slopes in Equation 3.23.

These natural strains cannot be used directly to evaluate the strain energy because they 'overlap' and must first be transformed to obtain the Cartesian strains. The natural strains are related to the Cartesian strains by a matrix transformation corresponding to the equations for Mohr's circle (Crandall et al. 1978), that is

$$\{\varepsilon_N\} = \begin{bmatrix} \varepsilon_a \\ \varepsilon_b \\ \varepsilon_c \end{bmatrix} \begin{bmatrix} c_{ax}^2 & c_{ay}^2 & \sqrt{2}\,c_{ax}c_{ay} \\ c_{bx}^2 & c_{by}^2 & \sqrt{2}\,c_{bx}c_{by} \\ c_{cx}^2 & c_{cy}^2 & \sqrt{2}\,c_{cx}c_{cy} \end{bmatrix} \begin{Bmatrix} \varepsilon_x \\ \varepsilon_y \\ \varepsilon_{xy} \end{Bmatrix} = C_N\{\varepsilon_C\} \tag{3.46}$$

Then the element is formulated by using interpolation functions such as the cubic functions of Equations 3.29 and differentiating these according to Equations 3.45 to form a strain interpolation matrix B_N for the natural strains. This is transformed to obtain the Cartesian strains using

$$\{\varepsilon_C\} = C_N^{-1} B_N \{d_N\} = C_N^{-1} B_N T\{d\} = B T\{d\} \tag{3.47}$$

where T is a matrix like that in Equation 3.23 which, transforms the global freedoms to local natural values.

Then the element stiffness matrix is given by

$$k = T^t \left(\int B^t D B \, dV \right) T \tag{3.48}$$

where D is the modulus matrix

$$D = Et/(1 - v^2)\begin{bmatrix} 1 & v & 0 \\ 0 & 1 & 0 \\ 0 & 0 & 1-v \end{bmatrix} \tag{3.49}$$

Note that, in order to calculate the strain energy correctly, the term $(1 - v)$ for the shear modulus in Equation 3.49 is twice the usual value because of the $\sqrt{2}$ factors used in Equation 3.46. Then in calculating strains using Equation 3.47 the $(1 - v)$ term in Equation 3.49 must be divided by $\sqrt{2}$.

The $\sqrt{2}$ factors in Equation 3.46 are used to ensure equivalence of this transformation to a two step transformation procedure used by Argyris (1968) and used in Section 8.2.

In thin plate elements the curvatures are calculated as generalized strains resulting from integration of the strains through the plate thickness, so that t in matrix D is replaced by $t^3/12$. Then again Equation 3.46 can be also used to transform from natural curvatures to Cartesian values.

As in Section 3.3 the global slopes are transformed to natural slopes using a matrix T and the interpolation formed in terms of these. Then the natural curvature interpolation matrix B_N is calculated from the interpolation functions using an extension of Equations 3.25 and 3.26 reminiscent of second order finite difference calculations

$$\begin{aligned}
\chi_a &= L_a(\partial \phi_a / \partial a) = \partial \phi_a / \partial L_2 - \partial \phi_a / \partial L_1 \\
&= \partial^2 w / \partial L_1^2 + \partial^2 w / \partial L_2^2 - 2\partial^2 w / \partial L_1 \partial L_2 \\
\chi_b &= \partial^2 w / \partial L_2^2 + \partial^2 w / \partial L_3^2 - 2\partial^2 w / \partial L_2 \partial L_3 \\
\chi_c &= \partial^2 w / \partial L_3^2 + \partial^2 w / \partial L_1^2 - 2\partial^2 w / \partial L_3 \partial L_1
\end{aligned} \tag{3.50}$$

and the element stiffness matrix is given by Equation 3.48 [with $B = C_N^{-1} B_N$].

3.7. References

Argyris JH, Three dimensional anisotropic and inhomogeneous media - matrix analysis for small and large displacements, *Ingenieur Archiv* 31 (1968) 33-49.

Crandall SH, Dahl NC, Lardner TJ, *An Introduction to the Mechanics of Solids*, 2nd edn, McGraw-Hill Kogakusha, Tokyo, 1978.

Ergatoudis J, Irons BM, Zienkiewicz OC, Curved isoparametric element for finite element analysis, *Int. J. Solids & Structures* 7 (1968) 31.

Mohr GA, *Finite Elements for Solids, Fluids, and Optimization*, Oxford University Press, Oxford, 1992.

Chapter 4

THIN PLATE ELEMENTS
USING BASIS TRANSFORMATION

4.1. The BCIZ element

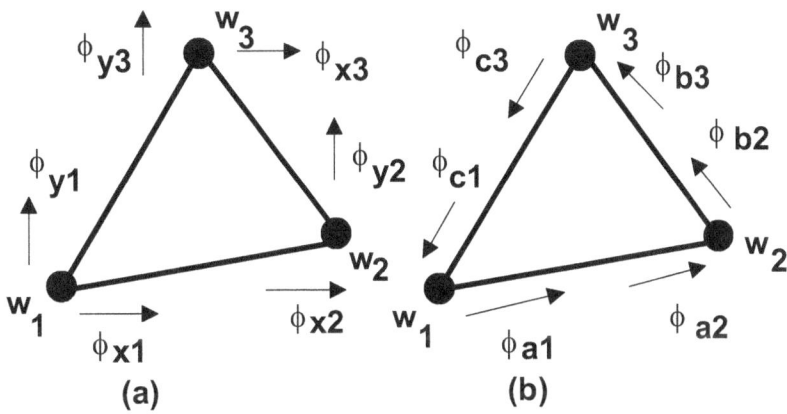

Figure 4.1. Nine freedom thin plate element:
(a) global freedoms, (b) local freedoms

The BCIZ element (Bazeley et al, 1965) was the first 'satisfactory' nine freedom thin plate element, other early solutions arbitrarily omitting one term from the full 10 term cubic expansion (Tocher, 1962) or combining three 'sub elements' (Clough & Tocher, 1962).

To obtain a concise formulation (Mohr, 1992) it is first convenient to transform the global slope ϕ_x, ϕ_y freedoms to the natural slopes shown in Figure 4.1(b) using the matrix T of Equation 3.23, excluding the inconvenient centroidal freedom by omitting the tenth row and column.

Then the interpolation functions for the BCIZ element are

$$f_1 = L_1 + L_1^2 L_2 + L_1^2 L_3 - L_2^2 L_1 - L_3^2 L_1 \quad \text{(for } w_1\text{)}$$

$$f_2 = L_2 + L_2^2 L_3 + L_2^2 L_1 - L_3^2 L_2 - L_1^2 L_2 \quad \text{(for } w_2\text{)}$$

$$f_3 = L_3 + L_3^2 L_1 + L_3^2 L_2 - L_1^2 L_3 - L_2^2 L_3 \quad \text{(for } w_3\text{)}$$

$$f_4 = L_1^2 L_2 + \frac{1}{2} L_1 L_2 L_3, \quad f_5 = -L_1^2 L_3 - \frac{1}{2} L_1 L_2 L_3 \quad \text{(for } L_a\phi_{a1}, L_a\phi_{c1}\text{)}$$

$$f_6 = L_2^2 L_3 + \frac{1}{2} L_1 L_2 L_3, \quad f_7 = -L_2^2 L_1 - \frac{1}{2} L_1 L_2 L_3 \quad \text{(for } L_b\phi_{b2}, L_b\phi_{a2}\text{)}$$

$$f_8 = L_3^2 L_1 + \frac{1}{2} L_1 L_2 L_3, \quad f_9 = -L_3^2 L_2 - \frac{1}{2} L_1 L_2 L_3 \quad \text{(for } L_c\phi_{c3}, L_c\phi_{b3}\text{)}$$

$$\text{(4.1)}$$

The interpolation functions are those of Equations 3.29 with the tenth function for the centroidal displacement omitted and the $L_1 L_2 L_3$ term removed from the first three functions and adjusted in the remaining functions. This term is sometimes referred to as the *bubble function* because it describes the shape of a triangular membrane subjected to uniformly distributed moments along its edges (Timoshenko & Woinowsky-Krieger, 1956). The adjustments of the bubble function terms in the functions for the slopes were arrived at by seeking to satisfy condition that interpolations must, as a minimum, represent constant strain in the element exactly and Meek (1972) gives a derivation of the $\pm(1/2)$ terms in Equations 4.1.

Note too that the functions of Equations 4.1 can also be obtained from the standard Hermitian cubic interpolation functions of Equations 3.29 by using Equation 4.14 (Mohr, 1998).

Applying Equations 3.50 to these functions the interpolation matrix for the natural curvatures χ_a, χ_b, χ_c is obtained as

$$B_N^t = \begin{bmatrix} 2L_3 + 2b_1 & -4L_1 & 2L_2 - b_3 \\ 2L_3 - 2b_1 & 2L_1 + 2b_2 & -4L_2 \\ -4L_3 & 2L_1 - 2b_2 & 2L_2 + 2b_3 \\ b_1 - 1 & -L_1 & L_2 \\ -L_3 & L_1 & b_3 + 1 \\ L_3 & b_2 - 1 & -L_2 \\ b_1 + 1 & -L_1 & L_2 \\ -L_3 & L_1 & b_3 - 1 \\ L_3 & b_2 + 1 & -L_2 \end{bmatrix} \quad \text{(4.2)}$$

where

$$b_1 = 3(L_2 - L_1), \quad b_2 = 3(L_3 - L_2), \quad b_3 = 3(L_1 - L_3) \qquad (4.3)$$

and the rows must be divided by L_a^2, L_b^2, L_c^2 respectively.

Then using Equation 3.46 B_N is transformed to obtain the interpolation matrix for the Cartesian curvatures

$$B = C_N^{-1} B_N \qquad (4.4)$$

and the final element stiffness matrix is given by three point integration at the midsides as

$$k = T_9^t \left[\sum B^t D_p B (\Delta/3) \right] T_9 \qquad (4.5)$$

where T_9 denotes matrix T of Equation 3.23 with its tenth row and column omitted and D_P is the modulus matrix D of Equation 3.49 with t replaced by $t^2/12$ for the case of plate bending.

Formulation of the element is much simplified using natural strains and the simple explicit interpolations of Equations 4.1 are still much used.

4.2. CBTP element with cubic Lagrangian basis

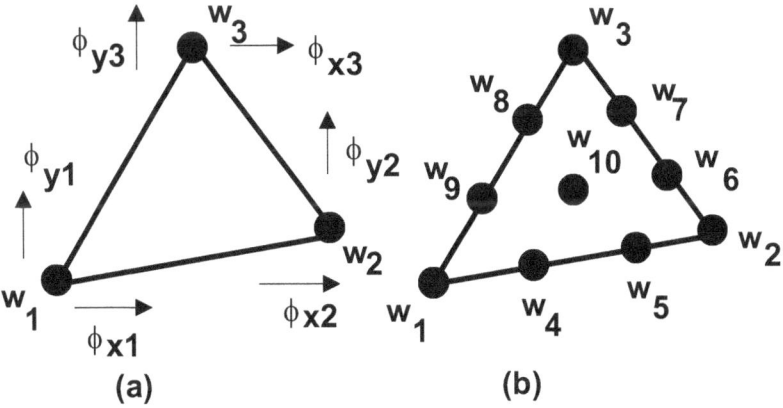

Figure 4.2. Nine freedom thin plate element: (a) global freedoms, (b) local freedoms

Figure 4.2(a) shows the usual global freedoms for a nine freedom thin plate element.

To establish a basis transformation to obtain the local freedoms shown in Figure 4.2(b) one dimensional cubic interpolation is applied on each side of the element to determine the values of w at fictitious 'local' nodes at the third points (Mohr and Mohr, 1986). On side 12, for example, the interpolation is

$$w = f_1 w_1 + f_2(L_a \phi_{a1}) + f_3 w_2 + f_4(L_a \phi_{a2}) \tag{4.6}$$

where

$$f_1 = 1 - 3s^2 + 2s^3, \quad f_2 = s - 2s^2 + s^3, \quad f_3 = 3s^2 - 2s^3. \quad f_4 = s^3 - s^2 \quad s = 0 \to 1 \tag{4.7}$$

and are the natural slopes parallel to this side and these are obtained using Equations 3.20 and 3.21, yielding

$$w = f_1 w_1 + f_2(x_{21}\phi_{x1} + y_{21}\phi_{y1}) + f_3 w_3 + f_4(x_{21}\phi_{x2} + y_{21}\phi_{y2}) \tag{4.8}$$

At nodes 4 and 5 we have $s = 1/3$ and $s = 2/3$ and substituting these values into Equations 4.7 and using the results in Equation 4.8

$$27 w_4 = 20w_1 + 4x_{21}\phi_{x1} + 4y_{21}\phi_{y1} + 7w_2 - 2x_{21}\phi_{x2} - 2y_{21}\phi_{y2} \tag{4.9}$$

$$27 w_5 = 7w_1 + 2x_{21}\phi_{x1} + 2y_{21}\phi_{y1} + 20w_2 - 4x_{21}\phi_{x2} - 4y_{21}\phi_{y2} \tag{4.10}$$

Repeating this exercise on the other two sides the required basis transformation is

$$\{d_N\} = \{w_1.w_2, w_3 \ldots w_{10}\} = T\{w_1, \phi_{x1}, \phi_{y1} \ldots w_3, \phi_{x3}, \phi_{y3}\} = T\{d\} \tag{4.11}$$

where

$$T = (1/27) \begin{bmatrix}
27 & 0 & 0 & 0 & 0 & 0 & 0 & 0 & 0 \\
0 & 0 & 0 & 27 & 0 & 0 & 0 & 0 & 0 \\
0 & 0 & 0 & 0 & 0 & 0 & 27 & 0 & 0 \\
20 & 4x_{21} & 4y_{21} & 7 & -2x_{21} & -2y_{21} & 0 & 0 & 0 \\
7 & 2x_{21} & 2y_{21} & 20 & -4x_{21} & -4y_{21} & 0 & 0 & 0 \\
0 & 0 & 0 & 20 & 4x_{32} & 4y_{32} & 7 & -2x_{32} & -2y_{32} \\
0 & 0 & 0 & 7 & 2x_{32} & 2y_{32} & 20 & -4x_{32} & -4y_{32} \\
7 & -2x_{13} & -2y_{13} & 0 & 0 & 0 & 20 & 4x_{13} & 4y_{13} \\
20 & -4x_{13} & -4y_{13} & 0 & 0 & 0 & 7 & 2x_{13} & 2y_{13} \\
& & \Sigma(\text{ rows } & 4\text{-}9 &)/4 - \Sigma & (\text{rows } & 1 - 3 &)/6 &
\end{bmatrix}$$

$$\tag{4.12}$$

The bottom row of this matrix uses the approximation

$$w_{10} = (w_4 + w_5 + w_6 + w_7 + w_8 + w_9)/4 - (w_1 + w_2 + w_3)/6 \tag{4.13}$$

To derive this we substitute the areal coordinates $(1/3, 1/3, 1/3)$ of the centroid into Equations 4.1 to obtain

$$w_{10} = (w_1 + w_2 + w_3)/3 + L_a(\phi_{a1} - \phi_{a2})/18 + L_b(\phi_{b2} - \phi_{b3})/18 + L_c(\phi_{c3} - \phi_{c1})/18 \tag{4.14}$$

Combining Equations 3.20, 4.9 and 4.10

$$w_4 = (20w_1 + 4L_a\phi_{a1} + 7w_2 - 2L_a\phi_{a2})/27$$
$$w_5 = (7w_1 + 2L_a\phi_{a1} + 20w_2 - 4L_a\phi_{a2})/27 \tag{4.15}$$

and adding these

$$L_a(\phi_{a1} - \phi_{a2}) = (27/6)(w_4 + w_5 - w_1 - w_2) \tag{4.16}$$

Cyclic progression gives the results for the other two element sides

$$L_b(\phi_{b2} - \phi_{b3}) = (27/6)(w_6 + w_7 - w_2 - w_3)$$
$$L_c(\phi_{c3} - \phi_{c1}) = (27/6)(w_8 + w_9 - w_3 - w_1) \tag{4.17}$$

Substituting Equations 4.15 - 4.17 into Equation 4.14 yields Equation 4.13, the bottom row of the basis transformation matrix T of Equation 4.12.

Applying this to the global freedoms of Figure 4.2(a) gives the local freedoms of Figure 4.2(b) and to these the standard cubic Lagrangian areal coordinate interpolation can be applied. These are Equations 3.17, but note that the node numbering of Figure 4.2(b) differs from that of Figure 3.3 and the interpolation functions are renumbered here accordingly:

$$f_1 = 4.5L_1^3 - 4.5L_1^2 + L_1, \quad f_2 = 4.5L_2^3 - 4.5L_2^2 + L_2, \quad f_3 = 4.5L_3^3 - 4.5L_3^2 + L_3$$
$$f_4 = 13.5L_1^2L_2 - 4.5L_1L_2, \quad f_5 = 13.5L_2^2L_1 - 4.5L_1L_2$$
$$f_6 = 13.5L_2^2L_3 - 4.5L_2L_3, \quad f_7 = 13.5L_3^2L_2 - 4.5L_2L_3$$
$$f_8 = 13.5L_3^2L_1 - 4.5L_3L_1, \quad f_9 = 13.5L_1^2L_3 - 4.5L_3L_1$$
$$f_{10} = 27L_1L_2L_3$$

$$\tag{4.18}$$

The interpolation matrix for the natural curvatures is obtained by applying Equations 3.50 to Equations 4.18 to obtain

$$
B_N^t = \begin{bmatrix}
27L_1 - 9 & 0 & 27L_1 - 9 \\
27L_2 - 9 & 27L_2 - 9 & 0 \\
0 & 27L_3 - 9 & 27L_3 - 9 \\
27L_2 - 54L_1 + 9 & 0 & 27L_2 \\
27L_1 - 54L_2 + 9 & 27L_1 & 0 \\
27L_3 & 27L_3 - 54L_2 + 9 & 0 \\
0 & 27L_2 - 54L_3 + 9 & 27L_2 \\
0 & 27L_1 & 27L_1 - 54L_3 + 9 \\
27L_3 & 0 & 27L_3 - 54L_1 + 9 \\
-54L_3 & -54L_1 & -54L_2
\end{bmatrix} \tag{4.19}
$$

where the rows of this matrix must be divided by L_a^2, L_b^2, L_c^2 respectively.

Then the final stiffness matrix is given by Equations 4.4 and 4.5.

Using the basis transformations and interpolations of this element and the BCIZ element for potential flow problems it was discovered that identical results were obtained (Mohr, 1998). The global interpolation functions for the present element are given by transposing matrix T of Equation 4.12

$$\{f^*\} = T^t \{f\} \tag{4.20}$$

so that we obtain for w_1

$$27f_1{}^* = 27f_1 + 20f_4 + 7f_3 + 7f_8 + 20f_9 + 9f_{10} \tag{4.21}$$

in which the summation of Equation 4.13 is used to obtain the last term.

Substituting from Equations 4.18 in Equation 4.21, after eliminating L_3 using the identity $L_1 + L_2 + L_3 = 1$ we obtain

$$f_1{}^* = -2L_1^2 + 3L_1^2 - 2L_1^2 L_2 - 2L_1 L_2^2 + 2L_1 L_2 \tag{4.22}$$

exactly the same result as obtained by eliminating L_3 from the first of Equations 4.1. Now using column 2 of T in Equation 4.12 we obtain for ϕ_{x1}

$$27f_2{}^* = 4x_{21} f_4 + 2x_{21} f_5 - 2x_{13} f_8 - 4x_{13} f_9 + (6x_{21} - 6x_{13})f_{10}/4 \tag{4.23}$$

again using the summation of Equation 4.13 to obtain the last term.

Substituting from Equations 4.18 in Equation 4.23 we obtain after a little manipulation

$$f_2^* = x_{21}(L_1^2 L_2 + \tfrac{1}{2}L_1 L_2 L_3) - x_{13}(L_1^2 L_3 - \tfrac{1}{2}L_1 L_2 L_3) \qquad (4.24)$$

which is equivalent to the fourth and fifth of Equations 4.1 and incorporates the transformations of the transpose of matrix T of Equation 3.23.

Hence the CBTP element is exactly equivalent to the BCIZ element. Formulation of both elements is quite simple but the CBTP element has the advantage of clearly incorporating elimination of the centroidal freedom in the last row of its basis transformation matrix. Then, if we wished, the centroidal freedom is quickly restored by 'zeroing' the tenth row of T and adding a tenth column with a last entry of 27.

4.3. QBTP element with dual quadratic Lagrangian basis

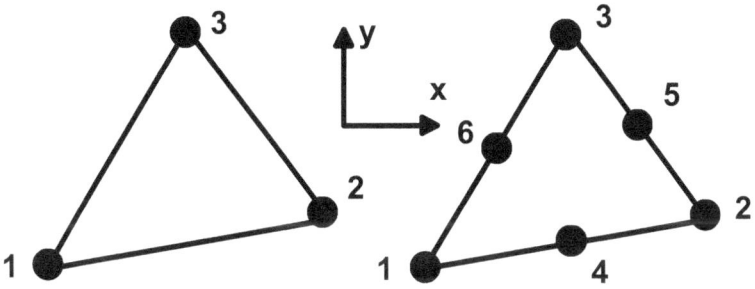

(a) w,ϕ_x,ϕ_y at each node   (b) ϕ_x, ϕ_y at each node

Figure 4.3. (a) Global freedoms, (b) local freedoms.

Figure 4.3(a) shows the usual nine freedom thin plate element freedoms and Figure 4.3(b) the local freedoms obtained by Mohr (1981) using the 'method of nested interpolations', the basis transformation method used in Section 4.2. The basis transformation required is achieved by applying cubic interpolation along each side of the element. Denoting slope parallel and normal to side 12 as ϕ_a and ϕ_{na} respectively, the parallel slope is given by differentiating the cubic interpolation

$$w = (1 - 3s^2 + 2s^3)w_1 + (s - 2s^2 + s^3)(L_a\phi_{a1}) + (3s^2 - 2s^3)w_2 + (s^3 - s^2)(L_a\phi_{a2})$$
$$(4.25)$$

$$L_a\phi_a = dw/ds = (6s^2 - 6s)w_1 + (1 - 4s + 3s^2)(L_a\phi_{a1}) + (6s - 6s^2)w_2 + (3s^2 - 2s)(L_a\phi_{a2})$$
$$(4.26)$$

Substituting $s = 0.5$

$$\phi_a = -1.5w_1/L_a - \phi_{a1}/4 + 1.5w_2/L_a - \phi_{a2}/4 \qquad (4.27)$$

Allowing only linear variation of the normal slope

$$\phi_{na4} = (\phi_{na1} + \phi_{na2})/2 \qquad (4.28)$$

In Equations 4.27 and 4.28 the parallel and normal slopes are given by

$$\phi_a = c_{ax}\phi_x + c_{ay}\phi_y, \quad \phi_{na} = -c_{ay}\phi_{ax} + c_{ax}\phi_y \qquad (4.29)$$

The required local freedoms at node 4 are given by the inverse of these

$$\phi_{x4} = c_{ax}\phi_{a4} - c_{ay}\phi_{na4}, \quad \phi_{y4} = c_{ay}\phi_{a4} + c_{ax}\phi_{na4} \qquad (4.30)$$

Substituting Equations 4.27, 4.28 and 4.29 into Equations 4.30 we obtain

$$\phi_{x4} = A_a(w_2 - w_1) + B_a\phi_{x1} + C_a\phi_{y1} + B_a\phi_{x2} + C_a\phi_{y2}$$
$$\phi_{y4} = D_a(w_2 - w_1) + C_a\phi_{x1} + E_a\phi_{y1} + C_a\phi_{x2} + E_a\phi_{y2} \qquad (4.31)$$

where

$$A_a = 1.5c_{ax}/L_a, B_a = (2c_{ay}^2 - c_{ax}^2)/4, \ C_a = -3c_{ax}c_{ay}/4, D_a = 1.5c_{ay}/L_a, \ E_a = (2c_{ax}^2 - c_{ay}^2)/4 \qquad (4.32)$$

Repeating this exercise for the other two sides the complete basis transformation matrix T for the local midside freedoms is given by

$$\begin{Bmatrix} \phi_{x4} \\ \phi_{y4} \\ \phi_{x5} \\ \phi_{y5} \\ \phi_{x6} \\ \phi_{y6} \end{Bmatrix} = \begin{bmatrix} -A_a & B_a & C_a & A_a & B_a & C_a & 0 & 0 & 0 \\ -D_a & C_a & E_a & D_a & C_a & E_a & 0 & 0 & 0 \\ 0 & 0 & 0 & -A_b & B_b & C_b & A_b & B_b & C_b \\ 0 & 0 & 0 & -D_b & C_b & E_b & D_b & C_b & E_b \\ A_c & B_c & C_c & 0 & 0 & 0 & -A_c & B_c & C_c \\ D_c & C_c & E_c & 0 & 0 & 0 & -D_c & C_c & E_c \end{bmatrix} = T\{d\} \qquad (4.33)$$

or

$$\{\phi_{xj}, \phi_{yj}\} = T\{w_i, \phi_{xi}, \phi_{yi}\} \quad i = 1 \to 3, \ j = 3 + i \qquad (4.34)$$

where A_b, B_b etc. are given by cyclic progression of subscripts a to b and c in Equations 4.32. The interpolation for the local freedoms is then simply

$$\phi_x = \Sigma\, f_i\, \phi_{xi}, \quad \phi_y = \Sigma\, f_i\, \phi_{yi} \quad i = 1 \rightarrow 6 \tag{4.35}$$

where

$$f_1 = 2L_1^2 - 1, \quad f_2 = 2L_2^2 - 1, \quad f_3 = 2L_3^2 - 1, \quad f_4 = 4L_1 L_2, \quad f_5 = 4L_2 L_3, \quad f_6 = 4L_3 L_1 \tag{4.36}$$

Defining the curvatures as

$$\chi_x = \partial\phi_x/\partial x, \quad \chi_y = \partial\phi_y/\partial y, \quad \chi_{xy} = \partial\phi_x/\partial y + \partial\phi_y/\partial x \tag{4.37}$$

formulation of the kernel stiffness matrix (for the local freedoms) is as for the classical linear strain triangle in Section 3.4, except that we substitute $t^3/12$ for t in the modulus matrix D.

Thus the final element stiffness matrix is given by

$$k = T^{*t}\, k^*\, T^* \quad \text{where} \quad k^* = \Sigma\, B^t DB\,(\Delta/3) \tag{4.38}$$

where T^* is the matrix T of Equation 4.33 augmented by a 6×9 Boolean matrix (placed first) corresponding to the vertex freedoms ϕ_{xi}, ϕ_{yi} $i = 1 \rightarrow 3$.

A 'stress matrix' = DBT is saved at each integration point in Equation 4.38, here for the midside nodes, and three further loops for calculation of stress matrices for the corner nodes can also be included. These matrices are used to calculate average nodal moments, that is the moments are averaged between elements impinging at a node.

Results with this element can be slightly improved by applying a small penalty factor to the constraint

$$\delta = \partial\phi_x/\partial y - \partial\phi_y/\partial x = 0 \tag{4.39}$$

so that the constrained element stiffness matrix is given by

$$k_r = k + \beta\, \Sigma\, P^t P(\Delta/3) \tag{4.40}$$

where P is a 1×12 constraint interpolation matrix formed according to Equation 4.39 using the matrix G of Equation 3.36.

The modified element was tested exhaustively on quadrants of square simply supported and clamped plates with blanket and central point loads and parameters $E = 12$, $\upsilon = 0$, $t = 1$ and L (plate span) $= 1$. $\beta = 10$ to $10{,}000$ gave 'stiff' solutions but it was found that excellent results were obtained using $\beta = \frac{1}{2}$ for all meshes. Indeed with this modification the element is clearly equal or superior in accuracy to complex quartic elements.

The reason for this accuracy is apparently a relaxation of the constraint of Equation 4.39.

Indeed $\beta = \frac{1}{2}$ corresponded to the twist modulus ($D_{XY} = Et^3/24(1-\upsilon)$) with the parameters used in the test problems and this is the value recommended for general purposes. Thus Equation 4.39 is an error term for the twisting curvature and, indeed, might be regarded as a fourth generalized strain when $\beta = D_{XY}$.

4.4. Numerical results

The element of Section 4.3 is a good deal more accurate than the elements of Sections 4.1 and 4.2 (these two being equivalent). The Lagrangian basis for this element does not, however, permit calculation of 'consistent' loads for blanket transverse loading (Mohr, 1998b). As an approximation, therefore, the consistent loads for the interpolation of Equations 4.1 can be used for all three elements.

In elements derived using basis transformation the consistent loads are calculated as

$$\{q_c\} = T^t \iint \{f\}\,dxdy = T^t \{q_c{}^*\} \tag{4.41}$$

and the (local) interpolation functions of Equations 4.1 can be exactly integrated using the formula

$$\iint L_1^a L_2^b L_3^c \, dxdy = [a!\,b!\,c!/(a+b+c+2)!](2\Delta) \tag{4.42}$$

yielding

$$\{q_c{}^*\} = \{8\,,8,8,1,-1,1,-1,1,-1\}\Delta/24 \tag{4.43}$$

and using the T matrix of Equation 3.23 the global consistent loads are

$$q_{c1} = q_{c4} = q_{c7} = q\Delta/3$$

$$q_{c2} = q\Delta(x_{21} - x_{13})/24, \quad q_{c3} = q\Delta(y_{21} - y_{13})/24$$

$$q_{c5} = q\Delta(x_{32} - x_{21})/24, \quad q_{c6} = q\Delta(y_{32} - y_{21})/24 \tag{4.44}$$

$$q_{c8} = q\Delta(x_{13} - x_{32})/24, \quad q_{c2} = q\Delta(y_{13} - y_{32})/24$$

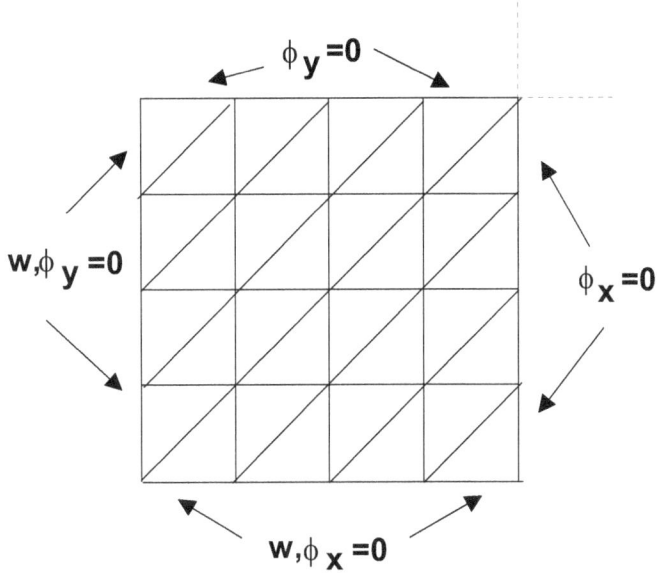

Figure 4.4.
Boundary conditions for quadrant of a simply supported square plate.

Figure 4.4 shows the mesh type used to test the three plate elements considered. For a clamped plate the boundary conditions are w, ϕ_x, $\phi_y = 0$ on both the supported edges and note that the diagonals are oriented to avoid a 'dead' element in the corner.

Table 4.1 shows results with the BCIZ element and the quadratic basis QBTP element, demonstrating the much greater accuracy of the QBTP element and that this is improved slightly using $\beta = \frac{1}{2}$ and consistent loads.

Table 4.1 also includes results obtained with an 18 freedom quintic triangular element (Mohr, 1992). This uses 6 df per node so that here the number of freedoms used per quadrant is divided by 3 to obtain an 'equivalent' number of nodes. With practical numbers of nodes/elements the QBTP element is almost as accurate as the highly accurate quintic element.

Table 4.1. Results for blanket loaded clamped plate ($v = 0$).

Nodes	$10^5 w_C/qL^4$	$10^5 m_C/qL^2$	$-10^5 m_E/qL^2$
BCIZ			
9	1560	1431	5480
25	1350	1777	5530
81	1290	1838	5450
QBTP, $\beta = 0$			
9	1537	2359	5283
25	1344	1939	5299
49	1301	1843	5243
81	1286	1809	5213
QBTP, $\beta = 1/2$			
9	1505	2338	5082
25	1334	1931	5175
49	1296	1840	5157
81	1283	1807	5148
QBTP, $\beta = 1/2$, cons. loads			
9	1422	2203	5016
25	1307	1867	5154
49	1284	1804	5148
81	1276	1784	5143
Quintic,			
8 (equiv.)	1149	1736	3926
18	1264	1765	4964
32	1265	1762	5102
50	1265	1762	5123
Exact	1265	1777	5135

Finally Table 4.2 shows results with the QBTP element for simply supported plates with blanket and central point loads. Again the results are excellent and results of similar accuracy are also obtained for the case of a clamped plate with central load (Mohr, 1998b).

Table 4.2. QBTP element results for simply supported plates, $\beta = 1/2$

Nodes	Blanket load (cons. loads)		Central load	
	$10^6{}_C/qL^4$	$10^5{}_C/qL^2$	$10^6 w_C/qL^4$	$10^4 m_C/qL^2$
4	3968	4762	11905	1429
9	4012	3927	11532	1900
16	4043	3808	11574	2230
25	4053	3750	11590	2464
36	4057	3726	11596	2645
49	4059	3712	11598	2791
64	4059	3704	11598	2915
81	4060	3700	11601	3023
Exact	4062	3684	11601	3100

4.5. Conclusions

The classical BCIZ element is easy to derive and is still much used. Its convergence is often a little erratic, however, and the quadratic basis QBTP element is much more accurate.

That the cubic basis CBTP element is in fact equivalent to the BCIZ element is surprising indeed as the two formulations look completely different. The CBTP formulation is equally simple to formulate and has the advantage that when a centroidal node is required it is very easy to include because the element has a local centroidal freedom.

The BCIZ and CBTP elements both provide useful examples of basis transformation and the use of natural strains.

Finally, however, the quadratic basis QBTP element is very accurate, being almost as accurate as a quintic element with twice as many freedoms. Use of a small penalty factor of magnitude equal to the twist modulus slightly improves results, as do approximate consistent loads. It is therefore expected that this recently improved element will be widely used in future.

4.6. References

Bazeley GP, Cheung YK, Irons BM, Zienkiewicz OC, Triangular elements in plate bending - conforming and nonconforming solutions, *Proc. Conf. Matrix Methods in Structural Mechanics*, Wright-Patterson Air Force Base, Ohio 1965.

Clough RW, Tocher LJ, Finite element stiffness matrices for analysis of plates in bending, *Proc. Conf. Matrix Methods in Structural Mechanics*, Wright-Patterson Air Force Base, Ohio 1965.

Meek JL, *Matrix Structural Analysis*, McGraw-Hill Kogakusha, Tokyo 1971.

Mohr GA, *Finite Elements for Solids, Fluids, and Optimization*, Oxford University Press, Oxford 1992.

Mohr GA, Mohr RS, A new thin plate finite element by basis transformation, *Computers & Structures* 22 (1986) 239.

Mohr GA, On two equivalent thin plate elements, *Communications in Numerical Methods in Engineering* 14 (1998) 271.

Mohr GA, Improving an accurate thin plate element, *Computer Methods in Applied Mechanics and Engineering* 166 (1998) 341.

Timoshenko SP, Woinowsky-Krieger WK, *Theory of Plates and Shells*, 3rd edn, McGraw-Hill, New York, 1956.

Tocher JL, *Analysis of Plate Bending Using Triangular Elements*, PhD thesis, Civil Engineering Dept, University of California, Berkeley, 1962.

Chapter 5

PLANE STRESS ELEMENTS
USING BASIS TRANSFORMATION

5.1. The drilling freedom

As a minimum plane stress elements, like the LST element of Section 3.4, must have translational freedoms *u*, v at each node and thin plate elements, like those of Chapter 4, usually have freedoms w, ϕ_x, ϕ_y.

Combining such elements for form *facet* elements for the analysis of shell structures we must include a sixth *drilling freedom* ϕ_z to permit coordinate transformation in three dimensions without components of the other rotations being lost on transformation.

Another problem without the drilling freedom included is that there are only five equations for each node. Coordinate transformation will give six equations for each node but the transformed global element stiffness matrix will still be of rank five. Hence if all elements meeting at a node are coplanar a singularity will be introduced into the structure stiffness matrix.

An alternative approach which does not require the drilling freedom is to assemble the equations in local coordinates. This is the usual procedure in curved shell elements but with flat shell elements it involves difficulties in that the local coordinates abruptly change at nodes, exactly where a unique definition is needed.

Thus, rather than 'waste' the drilling freedom it can be included in the plane stress part of the formulation by defining it as

$$\phi_z = \partial u/\partial y - \partial v/\partial x \tag{5.1}$$

The drilling freedom is logically included in a twelve freedom rectangular plane stress element by Mohr (1981), allowing complete quadratic interpolations and yielding marginally better results than the eight freedom bilinear element in applications to plane stress and folded plate problems. The element cannot be generalized into quadrilateral shape, however, and is thus of limited application.

51

Tocher and Hartz (1967) derived an eighteen freedom triangular plane stress element by allowing the freedoms

$$u, v, \partial u/\partial x, \partial v/\partial y, \gamma_{xy} = \partial u/\partial y + \partial v/\partial x, \phi_z$$

at each vertex. The element was formed by applying cubic interpolations to u and v in three sub-triangles of the element, so that condensation from a total of fifty four freedoms to the final eighteen must be achieved by matrix transformation.

A more economical element, the hybrid equivalent of which was first obtained by Dungar and Severn (1969), is used by Olson and Bearden (1979) to form a flat shell element. It is based on an incomplete nine term cubic displacement function proposed by Holand (1969) which has freedoms $u, v, \partial u/\partial x, \partial u/\partial y, \partial v/\partial x, \partial v/\partial y$, giving a total of 18 freedoms which is reduced to nine by applying various constraints.

This element yielded little improvement over using the constant strain triangle (with small artificial stiffnesses for the drilling freedom). In addition the solution was not explicit and required inversion of a 12×12 matrix.

There have, however, been relatively few elements incorporating the drilling freedom developed, other examples including those of Bergan and Felippa (1985), Taylor and Simo (1985), and Stander and Wilson (1989).

In this chapter, therefore, two simple and economical nine freedom plane stress elements which use *natural* definitions of the drilling freedom and basis transformation are derived.

5.2. The DFT1 element

To establish a natural definition of the drilling freedom the natural extensional freedoms α, β, γ parallel to each side shown in Figure 5.1 are defined by

$$\alpha = c_{ax}u + c_{ay}v, \quad \beta = c_{bx}u + c_{by}v, \quad \gamma = c_{cx}u + c_{cy}v \tag{5.2}$$

where the direction cosines are defined by the permutations of Equations 3.21, that is

$$c_{ax} = x_{21}/L_a, \ c_{ay} = y_{21}/L_a, \ c_{bx} = x_{32}/L_{ba}, \ c_{by} = y_{32}/L_b, \ c_{cx} = x_{13}/L_c, \ c_{cy} = y_{13}/L_c \tag{5.3}$$

Calculating the 'cross derivatives' at node 1

$$\partial \alpha/\partial c = (\partial \alpha/\partial x)(\partial x/\partial c) + (\partial \alpha/\partial y)(\partial y/\partial c) = c_{cx}(\partial \alpha/\partial x) + c_{cy}(\partial \alpha/\partial y)$$

$$\partial \gamma/\partial a = (\partial \gamma/\partial x)(\partial x/\partial a) + (\partial \gamma/\partial y)(\partial y/\partial a) = c_{ax}(\partial \gamma/\partial x) + c_{ay}(\partial \gamma/\partial y) \tag{5.4}$$

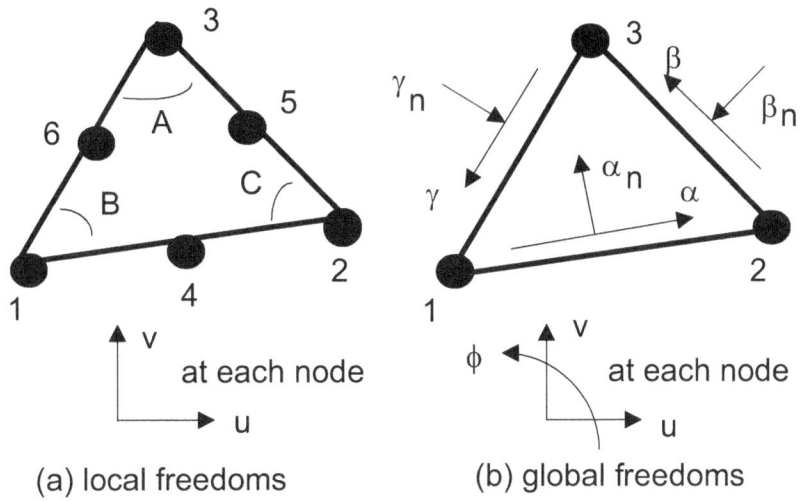

Figure 5.1. DFT1 element, showing natural coordinates in (b).

Combining Equations 5.2 - 5.4

$$\partial a/\partial c - \partial\gamma/\partial a = \{\partial u/\partial y - \partial v/\partial x\}[x_{13}y_{21} - x_{21}y_{13}]/L_aL_b = 4\Delta\phi_{z1}/L_aL_b \tag{5.5}$$

It follows by cyclic progression that the natural drilling freedoms at the vertices are given by

$$\partial a/\partial c - \partial\gamma/\partial a = 4\Delta\phi_{z1}/L_aL_b$$

$$\partial\beta/\partial a - \partial a/\partial b = 4\Delta\phi_{z2}/L_bL_c \tag{5.6}$$

$$\partial\gamma/\partial b - \partial\beta/\partial c = 4\Delta\phi_{z3}/L_cL_a$$

where $\phi_{z1}, \phi_{z2}, \phi_{z3}$ are the Cartesian drilling freedoms to be included at the vertices to obtain a nine freedom element.

Applying quadratic interpolation to the natural displacement α on side 31 using

$$a = (1 - 3s + 2s^2)a_1 + (4s - 4s^2)a_6 + (2s^2 - s)a_3 \qquad s = 0 \to 1 \tag{5.7}$$

and differentiating with respect to the natural coordinate c on this side

$$\partial a/\partial c = (\partial a/\partial s)(\partial s/c) = [(4s - 3)a_3 + (4 - 8s)a_6 + (4s - 1)a_1]/L_c \tag{5.8}$$

Substituting $s = 1$ (node 1) gives

$$(\partial a/\partial c)_1 = [a_3 - 4a_6 + 3a_1]/L_c \qquad (5.9)$$

Evaluating the other terms on the left sides of Equations 5.6 in like fashion gives

$$4\Delta\phi_{z1} = L_a(a_3 + 3a_1) + L_c(3\gamma_1 + \gamma_2) - 4L_a a_6 - 4L_c\gamma_4$$
$$4\Delta\phi_{z2} = L_b(\beta_1 + 3\beta_2) + L_a(3a_2 + a_3) - 4L_b\beta_4 - 4L_a a_5 \qquad (5.10)$$
$$4\Delta\phi_{z3} = L_c(\gamma_2 + 3\gamma_3) + L_b(3\beta_3 + \beta_1) - 4L_c\gamma_5 - 4L_b\beta_6$$

On side 31 we have

$$a_6 = c_{ax}U_6 + c_{ay}V_6 \qquad (5.11a)$$

where

$$U_6 = c_{cx}\gamma_6 - c_{cy}\gamma_{n6}, \quad V_6 = c_{cy}\gamma_6 + c_{cx}\gamma_{n6} \qquad (5.11b)$$

where γ_{n6} is the normal displacement at node 6, and constraining the parallel displacement on each side to linear variation

$$\gamma_6 = (\gamma_3 + \gamma_1)/2 = (c_{cx}U_3 + c_{cy}V_3 + c_{cx}U_1 + c_{cy}V_1)/2 \qquad (5.12)$$

and substituting Equations 5.11 and 5.12 in the first of Equations 5.10 a_6 is replaced by γ_{n6}. In like fashion β_6 in the third of Equations 5.10 is also replaced by γ_{n6}.

Thus three equations are obtained in terms of the normal slopes at the midsides and the global vertex displacements (including the drilling freedoms). Owing to the cyclic nature of the equations used to derive them, however, these three equations are not independent (Mohr, 1982) and cannot be solved to obtain the normal midside slopes, as intended.

Alternatively, defining A,B,C as the included angles at the element vertices shown in Figure 4.1(a), the natural displacements at node 1 are written as

$$\gamma = -a_n \sin B - a \cos B \quad \text{(on side12)}$$
$$a = \gamma_n \sin B - \gamma \cos B \quad \text{(on side 31)} \qquad (5.13)$$

Substituting Equations 5.13 in the first of Equations 5.6

$$4\Delta\phi_{z1}/L_aL_b = (\partial\gamma_n/\partial c)\sin B - (\partial\gamma/\partial c)\cos B + (\partial a_n/\partial a)\sin B + (\partial a/\partial a)\cos B \tag{5.14}$$

Constraining the angle at node 1 to remain constant and assuming linear variation of the parallel displacements the rotation at node 1 is given as

$$\psi_1 = (\partial\gamma_n/\partial c)_1 = (\partial a_n/\partial a)_1$$

$$= [4\Delta\phi_{z1}/L_aL_b + (\gamma_1 - \gamma_3)\cos B/L_c - (a_2 - a_1)\cos B/L_a]/2\sin B \tag{5.15}$$

and ψ_2, ψ_3 for the other nodes are obtained in like fashion.

Then on side 12 using Equation 4.25 and substituting $s = 0.5$ gives

$$a_{n4} = (a_{n1} + a_{n2})/2 + L_a(\psi_1 - \psi_2)/8 \tag{5.16}$$

and substituting this and

$$a_4 = (a_1 + a_2)/2 \tag{5.17a}$$

$$a_n = C_{ay}U - C_{ax}V \tag{5.17b}$$

into the 'reverse' transformation $U_4 = C_{ax}a_4 - C_{ay}a_{n4}$ we obtain after cancellations and noting that $c_{ax}^2 + c_{ay}^2 = 1$,

$$U_4 = (U_1 + U_2)/2 + C_{ay}L_a(\psi_2 - \psi_1)/8 \tag{5.18}$$

Similarly

$$V_4 = (V_1 + V_2)/2 + C_{ax}L_a(\psi_1 - \psi_2)/8 \tag{5.19}$$

The equations for U_5, V_5, U_6, V_6 follow by cyclic progression so that the required basis transformation is obtained as

$$
\begin{Bmatrix} U_4 \\ V_4 \\ U_5 \\ V_5 \\ U_6 \\ V_6 \end{Bmatrix} = (1/2)
\begin{bmatrix}
1 & 0 & 1 & 0 & 0 & 0 \\
0 & 1 & 0 & 1 & 0 & 0 \\
0 & 0 & 1 & 0 & 1 & 0 \\
0 & 0 & 0 & 1 & 0 & 1 \\
1 & 0 & 0 & 0 & 1 & 0 \\
0 & 1 & 0 & 0 & 0 & 1
\end{bmatrix}
\begin{Bmatrix} U_1 \\ V_1 \\ U_2 \\ V_2 \\ U_3 \\ V_3 \end{Bmatrix} + (1/8)
\begin{bmatrix}
-y_{21} & y_{21} & 0 \\
x_{21} & -x_{21} & 0 \\
0 & -y_{32} & y_{32} \\
0 & x_{32} & -x_{32} \\
y_{13} & 0 & -y_{13} \\
-x_{13} & 0 & x_{13}
\end{bmatrix}
\begin{Bmatrix} \psi_1 \\ \psi_2 \\ \psi_3 \end{Bmatrix}
$$

$$\tag{5.20}$$

where the rotations ψ_1, ψ_2, ψ_3 are given by the three permutations of Equation 5.15, transforming the natural displacements to Cartesian values using Equations 5.2,

$$
\left\{ \begin{array}{c} (2L_aL_c\tan B)\psi_1 \\ (2L_bL_a\tan C)\psi_2 \\ (2L_cL_b\tan B)\psi_3 \end{array} \right\} =
\left[\begin{array}{ccc}
L_cC_{ax}+L_aC_{cx} & -L_bC_{ax} & -L_bC_{cx} \\
L_cC_{ay}+L_aC_{cy} & -L_bC_{ay} & -L_bC_{cy} \\
4\Delta/\cos B & 0 & 0 \\
-L_cC_{ax} & L_aC_{bx}+L_bC_{ax} & -L_cC_{bx} \\
-L_cC_{ay} & L_aC_{by}+L_bC_{ay} & -L_cC_{by} \\
0 & 4\Delta/\cos C & 0 \\
-L_aC_{cx} & -L_aC_{bx} & L_bC_{cx}+L_cC_{bx} \\
-L_aC_{cy} & -L_aC_{by} & L_bC_{cy}+L_cC_{cy} \\
0 & 0 & 4\Delta/\cos A
\end{array} \right]^t
\left\{ \begin{array}{c} u_1 \\ v_1 \\ \phi_1 \\ u_2 \\ v_2 \\ \phi_2 \\ u_3 \\ v_3 \\ \phi_3 \end{array} \right\} = G^t\{d\}
$$

$$(5.21)$$

and multiplying this equation through by the cosines, then dividing the rows of G by the sines avoids division by the cosines, one of which is zero in right angled triangles.

The kernel stiffness matrix for the element is exactly that for the LST element of Section 3.4 and the final stiffness matrix is given by using the transformations of Equations 5.20 and 5.21.

5.3. The DFT2 element

To obtain an improved element it is simply assumed that on side 12 the natural derivatives of the transverse natural displacements at each end are equal to the vertex drilling freedoms, that is on side 12 we have

$$(\partial a_n/\partial a)_1 = \phi_{z1}, \quad (\partial a_n/\partial a)_2 = \phi_{z2}$$

$$(5.22)$$

Then applying cubic interpolation to the transverse natural displacement the value at the middle of this side is given by using Equation 5.16

$$a_n = (a_{n1} + a_{n2})/2 + s_a(\phi_{z1} - \phi_{z2})/8$$

$$(5.23)$$

Then combining the permutations of Equations (5.2), (5.17), (5.22) and (5.23) we obtain

$$
\begin{Bmatrix} a_4 \\ a_{n4} \\ \beta_5 \\ \beta_{n5} \\ \gamma_6 \\ \gamma_{n6} \end{Bmatrix} =
\begin{bmatrix}
c_{ax}/2 & c_{ay}/2 & 0 & c_{ax}/2 & c_{ay}/2 & 0 & 0 & 0 & 0 \\
-c_{ay}/2 & c_{ax}/2 & s_a/8 & -c_{ay}/2 & c_{ax}/2 & -s_a/8 & 0 & 0 & 0 \\
0 & 0 & 0 & c_{bx}/2 & c_{by}/2 & 0 & c_{bx}/2 & c_{by}/2 & 0 \\
0 & 0 & 0 & -c_{by}/2 & c_{bx}/2 & s_b/8 & -c_{by}/2 & c_{bx}/2 & -s_b/8 \\
c_{cx}/2 & c_{cy}/2 & 0 & 0 & 0 & 0 & c_{cx}/2 & c_{cy}/2 & 0 \\
-c_{cy}/2 & c_{cx}/2 & -s_c/8 & 0 & 0 & 0 & -c_{cy}/2 & c_{cx}/2 & s_c/8
\end{bmatrix} \{d\} = H\{d\}
$$

(5.24)

and transforming the midside natural displacements to Cartesian values using the permutations of Equations 5.11b

$$
\begin{Bmatrix} u_4 \\ v_4 \\ u_5 \\ v_5 \\ u_6 \\ v_6 \end{Bmatrix} =
\begin{bmatrix}
c_{ax} & -c_{ay} & 0 & 0 & 0 & 0 \\
c_{ay} & c_{ax} & 0 & 0 & 0 & 0 \\
0 & 0 & c_{bx} & -c_{by} & 0 & 0 \\
0 & 0 & c_{by} & c_{ax} & 0 & 0 \\
0 & 0 & 0 & 0 & c_{cx} & -c_{cy} \\
0 & 0 & 0 & 0 & c_{cy} & c_{cx}
\end{bmatrix}
\begin{Bmatrix} a_4 \\ a_{n4} \\ \beta_5 \\ \beta_{n5} \\ \gamma_6 \\ \gamma_{n6} \end{Bmatrix} = G\{d\}
$$

(5.25)

the complete displacement transformation is then given by

$$
\{u_1, v_1, - - u_6, v_6\} = \begin{bmatrix} C \\ GH \end{bmatrix} \{d\} = T\{d\}
$$

(5.26)

where

$$
C = \begin{bmatrix}
1 & 0 & 0 & 0 & 0 & 0 & 0 & 0 & 0 \\
0 & 1 & 0 & 0 & 0 & 0 & 0 & 0 & 0 \\
0 & 0 & 0 & 1 & 0 & 0 & 0 & 0 & 0 \\
0 & 0 & 0 & 0 & 1 & 0 & 0 & 0 & 0 \\
0 & 0 & 0 & 0 & 0 & 0 & 1 & 0 & 0 \\
0 & 0 & 0 & 0 & 0 & 0 & 0 & 1 & 0
\end{bmatrix}
$$

(5.27)

is a simple Boolean matrix for the translational vertex freedoms.
Then the final element stiffness matrix is given by the congruent transformation

$$
k_e = T^t k^* T
$$

(5.28)

where the kernel stiffness matrix $k*$ is exactly that for the linear strain triangle, and this is calculated by three point numerical integration at the midsid nodes in the present work.

Note in passing that omission of the six '$s/8$' entries in matrix H yields the CST (when small stiffnesses placed at (3,3), (6,6) and (9,9) in final stiffness matrix are needed to prevent singularity).

5.4. Numerical results

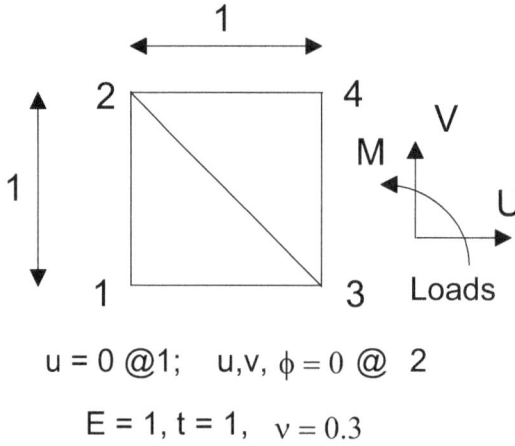

$$u = 0 \ @1; \quad u,v, \phi = 0 \ @ \ 2$$

$$E = 1, t = 1, \quad v = 0.3$$

Figure 5.2. Mesh for patch tests.

Figure 5.2 shows a two element patch test mesh. The three load cases used are:

(a) Constant shear:
$$V_1 = -0.5, U_3 = -0.5, V_3 = 0.5, U_4 = 0.5, V_4 = 0.5$$
(b) Constant direct stress:
$$M_1 = 1/12, M_2 = -1/12, U_3 = 0.5, M_3 = -1/12, U_4 = 0.5, M_4 = 1/12$$
(c) Pure bending:
$$M_1 = 0.1, M_2 = 0.1, U_3 = 1.2, M_3 = -0.1, U_4 = -1.2, M_4 = -0.1$$

The results for the DFT2 element are shown in Table 1, along with the exact results in parentheses. In case (a) the required unit shear stresses are obtained and in case (b) the required unit direct stresses are obtained. The constant stress tests are thus passed but note that, owing to the very simple manner in which ϕ_z is included in the formulation in Equation 5.22 , a datum value must be set, here zero at node 2, but that the solutions for the ϕ_z are nevertheless satisfactory in case (c).

Table 5.1. Solutions to patch tests of Fig. 5.2

Case	Node	u	v	f
(a)	1	0.0 (0)	0.0 (0)	0.0 (0)
	2	0.0 (0)	0.0 (0)	0.0 (0)
	3	0.0 (0)	2.6 (2.6)	0.0 (0)
	4	0.0 (0)	2.6 (2.6)	0.0 (0)
(b)	1	0.0 (0)	0.3 (0.3)	0.0 (0)
	2	0.0 (0)	0.0 (0)	0.0 (0)
	3	1.0 (1)	0.3 (0.3)	0.0 (0)
	4	1.0 (1)	0.0 (0)	0.0 (0)
(c)	1	0.0 (0)	0.3708 (0)	1.8446 (0)
	2	0.0 (0)	0.0 (0)	0.0 (0)
	3	4.0531 (6)	4.0531 (6)	11.5126 (12)
	4	-4.0531 (-6)	4.4239 (6)	9.6679 (12)

In contrast the element of Olson and Bearden (1979) gives poor results for (a) and (b) but approximate results for case (c). The DFT1 element, however, gives poor results for all three cases.

For case (c) the results are very similar to Olson and Bearden's results and only approximations. Accurate expectations for this case with only four nodes, however, are unrealistic with elements with incomplete quadratic interpolations (c.f. the LST direct stress results in Table 2) but when 8 elements (9 nodes) are used for this case, still setting $\phi_z = 0$ at node 2 (not a corner node now), the results are about 80% of the exact solutions, consistent with the 9 node displacement solution accuracy in Table 2. To put the present results into context note that though the LST satisfies tests (b) and (c) accurately, though needing 9 nodes of course, it gives very poor displacement results and only approximate stresses for test (a).

Fig. 5.3 shows the coarsest mesh for half a simply supported beam in which the load is uniformly distributed through the beam depth.

Table 5.2 shows the results obtained using the classical CST and LST elements, and the DFT1 and DFT2 elements. The stress results are average nodal stresses (note that, before averaging, the nodal stresses for the CST element are the same as those at its centroid).

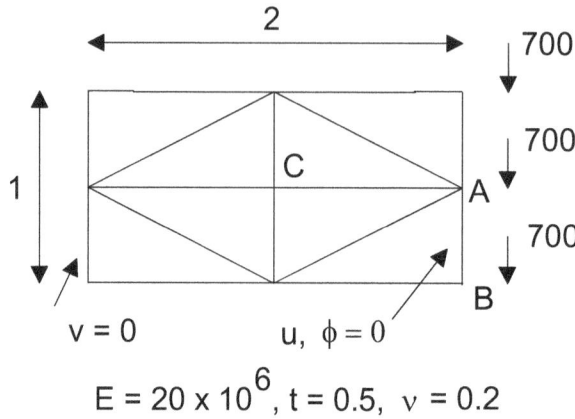

Figure 5.3. Coarse mesh for half a SS beam.

The two DFT elements give similar results which are almost as accurate as those of the LST (note that for the DFT1 element the displacement result for the 3x3 mesh with uni-directional diagonals was only 6.98, more consistent with expectations).

Table 5.2. Solutions for problem of Figure 5.3.

Element	Nodes	v_A	σ_B	τ_C
CST	3 x 3	3.510	7.16	1.181
	5 x 5	5.816	15.96	2.245
	7 x 7	6.828	20.26	2.681
LST	3 x 3	6.78	25.20	1.540
	5 x 5	7.77	25.20	3.675
	7 x 7	7.86	25.20	2.875
DFT1	3 x 3	7.287	17.29	3.430
	5 x 5	7.469	21.86	3.164
	7 x 7	7.686	23.84	3.148
DFT2	3 x 3	6.348 (6.4228)	16.12 (15.98)	3.086 (3.021)
( ) = no ϕ b.c	5 x 5	7.424 (7.5630)	21.78 (21.58)	3.153 (3.127)
at mid span	7 x 7	7.695 (7.7766)	23.87 (23.57)	3.151 (3.145)
Theory		8.000	25.20	3.150

With a complete quadratic interpolation the LST element is able to model the linear direct stress distribution exactly, whereas the DFT elements, of course, can only approximate it. The LST shear stress results, however, are clearly not convergent (Mohr, 1992), and the DFT elements rectify this problem.

Finally, note that direct moment loadings can also be used in problems such as that of Figure 5.3 but, owing to the approximations for ϕ_z, only with approximate results.

5.5. Conclusions

Because of their usefulness in flat shell elements for the analysis of shell structures plane stress elements including the drilling freedom are a subject of continuing research (Choi et al., 2002, Lee & Choi, 2002).

The DFT1 and DFT2 elements provide further examples of the use of natural variables, coordinates and derivatives and of basis transformation.

These two elements appear to be of similar accuracy in practical problems such as that of Figure 5.3, but the DFT1 element does very poorly on the patch tests of Figure 5.2, so that the DFT2 element is that recommended.

As it models shear behaviour much better than the classical LST element, as well as being designed for use in facet shell elements, it is expected that this element, or modifications of it, will be widely used.

5.6. References

Bergan PG, Felippa CA, A triangular membrane element with rotational degrees of freedom, *Computer Methods in Applied Mechanics & Engineering* 50 (1985) 25.

Choi CK, Lee WH, Transition membrane elements with drilling freedom for local mesh refinements, *Structural & Engineering Mechanics* 3 (1995) 75.

Kee TY, Choi CK, A new quadrilateral 5-node non-conforming membrane element with drilling DOF, *Structural & Engineering Mechanics* 14 (2002) 699.

Dungar R, Severn RT, Triangular finite elements of variable thickness and their application to plate and shell problems, *J. Strain Analysis* 4 (1969) 10.

Holand I, The finite element method in plane stress analysis, in *Finite Element Methods in Stress Analysis* (ed. I Holand, K Bell) Tapir, Trondheim 1969.

Mohr GA, A simple rectangular membrane element including the drilling freedom, *Computers & Structures* 13 (1981) 483.

Mohr GA, Finite element formulation by nested interpolations: application to the drilling freedom problem, *Computers & Structures* 15 (1982) 185.

Mohr GA, *Finite Elements for Solids, Fluids, and Optimization*, Oxford University Press, Oxford 1992.

Mohr GA, A new facet shell element, *Int. J. Arts & Sciences* 1 (2001) 36.

Olsen MD, Bearden TW, A simple triangular shell element revisited, *Int. J. Numerical Methods in Engineering* 14 (1979) 51.

Stander N, Wilson EL, A 4-node quadrilateral membrane element with in-plane vertex rotations and modified reduced quadrature, *Finite Elements in Analysis & Design* 6 (1989) 266.

Taylor RL, Simo JC, Bending and Membrane elements for analysis of thick and thin shells, *Proc. NUMETA 85 conference* (eds J Middleton, GN Pande), Swansea, 1985. Published by AA Balkema, Rotterdam.

Tocher JL, Hartz BJ, Higher order finite element for plane stress, *J. Engineering Mechanics Division, ASCE* (American Society of Civil Engineers), 93 (1967) 5402.

Chapter 6

FACET SHELL ELEMENTS

6.1. Flat triangular elements

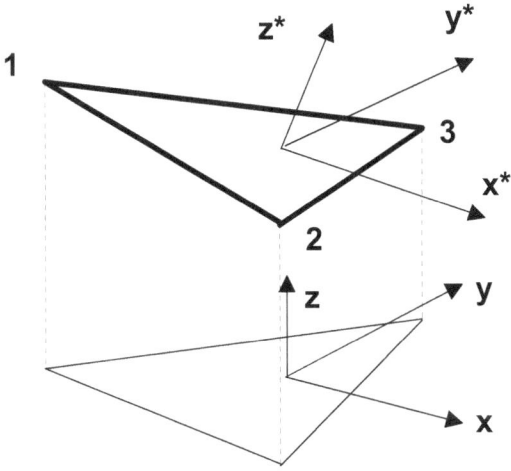

Figure 6.1. Flat triangular shell element.

The finite element method was originally developed by aeronautical engineers and analysis of shell structures was their aim. These are most easily modeled by combining flat triangular *facet* elements for plane stress and thin plate bending to form *facet* shell elements (Clough and Johnson, 1968; Parekh, 1969). To model the extensional behaviour we require only two in-plane translations and for the flexural behaviour a transverse deflection and two associated slopes. To permit coordinate transformation in three dimensions, however, we must have six freedoms $u, v, w, \phi_x, \phi_y, \phi_z$ because, without the drilling freedom ϕ_z, components of the other rotations would be lost on transformation.

The difficulty can be overcome by assigning small artificial stiffnesses to the drilling freedom (Zienkiewicz, 1977). Then, for example, the constant strain triangle of Section 3.1 can be used to model the plane stress action and the BCIZ element of Section 4.1 can be used to model the flexural behaviour.

Here we shall use the DFT2 element of Section 5.3 to incorporate the drilling freedom rigorously, together with the quadratic basis QBTP thin plate element of Section 4.3 to model the flexural behaviour. The local freedoms used by the two elements are respectively u^*, v^*, ϕ_z^* and w^*, ϕ_x^*, ϕ_y^* and after basis transformation their local freedoms are those shown in Figure 6.2.

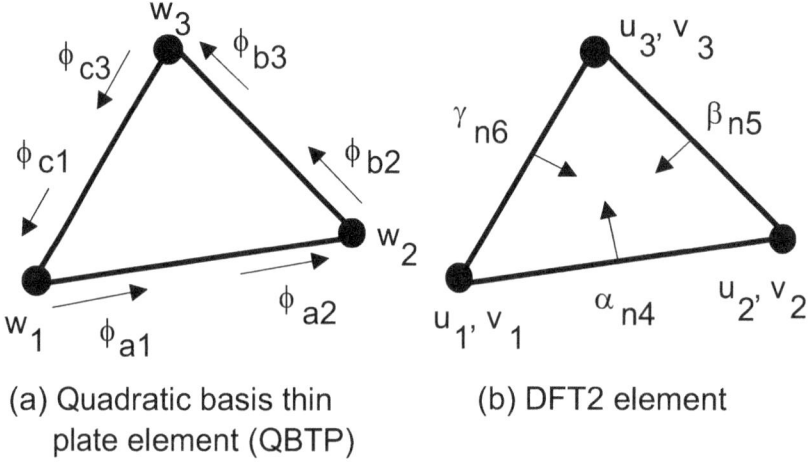

(a) Quadratic basis thin   (b) DFT2 element
    plate element (QBTP)

Figure 6.2. Local freedoms after basis transformation
for the component elements.

Note that in the DFT2 element the normal displacements shown, together with average parallel displacements, are finally transformed to u,v at the midsides.

Then the kernel local element stiffness matrix is given by

$$k^* = \begin{bmatrix} k_p & O_9 \\ O_9 & k_f \end{bmatrix}$$

(6.1)

where k_p is the stiffness matrix for the DFT2 element, k_f is the stiffness matrix for the QBTP element and O_9 is a 9×9 null matrix.

To obtain the transformation from the global to the local axes we begin by writing a linear interpolation for the elevation of the element surface

$$z = a + bx + cy$$

(6.2)

Substituting the nodal elevations and solving for b and c yields the solutions for the tangents of the slopes of the local x^* and y^* axes. These are the same as the solutions obtained for the coefficients of the Cartesian interpolation for the constant strain triangle in Section 3.1, that is, rows two and three in Equation 3.4

$$t_x = \partial z / \partial x = b = -(y_{32} z_1 + y_{13} z_2 + y_{21} z_3)/2\Delta \qquad (6.3)$$

$$t_y = \partial z / \partial y = c = (x_{32} z_1 + x_{13} z_2 + x_{21} z_3)/2\Delta \qquad (6.4)$$

where Δ is the area of the horizontal projection of the element.

Then the required coordinate transformation is (Mohr, 1976)

$$\left\{ \begin{array}{c} x^* \\ y^* \\ z^* \end{array} \right\} = \left[\begin{array}{ccc} c_x & -s_x t_y & s_x \\ -s_y t_x & c_y & s_y \\ -c_y s_x & -c_x s_y & c_x c_y \end{array} \right] \left\{ \begin{array}{c} x \\ y \\ z \end{array} \right\} = T_c \{x\} \qquad (6.5)$$

where

$$s_x = \sin(\tan^{-1} t_x), \quad c_x = \cos(\tan^{-1} t_x)$$
$$s_y = \sin(\tan^{-1} t_y), \quad c_y = \cos(\tan^{-1} t_y) \qquad (6.6)$$

The first row of the matrix t is obtained by applying the standard transformation

$$x^* = c_x x + s_x z_x \qquad (6.7)$$

for the rotation from the x to the x^* axis, where

$$z_x = z - t_y y \qquad (6.8)$$

allows for the rotation from the y to the y^* axis.

The second row is obtained in like fashion whilst the third is obtained as the direction cosines of the vector cross product of the two unit vectors parallel to the x^* and y^* axes. If $\hat{i}, \hat{j}, \hat{k}$ are unit vectors parallel to the global axes then these two unit vectors are

$$\hat{v}_x^* = c_x \hat{i} + 0\hat{j} + s_x \hat{k}, \quad \hat{v}_y^* = 0\hat{i} + c_y \hat{j} + s_y \hat{k} \qquad (6.9)$$

Then

$$\hat{v}_z{}^* = \hat{v}_x{}^* \times \hat{v}_y{}^* = -c_y s_x \hat{i} - c_x s_Y \hat{j} + c_x c_y \hat{k} \qquad (6.10)$$

is the unit vector parallel to the z^* axis and its components are the required direction cosines.

The final 18 x 18 global element stiffness matrix is given by

$$k = T_s^t k^* T_s$$

$$(6.11)$$

where k^* is given by Equation 6.1 and

$$T_s = \begin{bmatrix} T_n & O_6 & O_6 \\ O_6 & T_n & O_6 \\ O_6 & O_6 & T_n \end{bmatrix}, \quad T_n = \begin{bmatrix} T_c & O_3 \\ O_3 & T_c{}^* \end{bmatrix} \qquad (6.12a)$$

where

$$T_c{}^* = \begin{bmatrix} 0 & 1 & 0 \\ -1 & 0 & 0 \\ 0 & 0 & 1 \end{bmatrix} T_c \qquad (6.12b)$$

is a Boolean transformation required because globally the rotational freedoms are the right handed system

$$\phi_x = -\partial w/\partial y, \; \phi_y = \partial w/\partial x, \; \phi_z = \partial u/\partial y - \partial v/\partial x$$

whereas matrix k_f is derived using local freedoms

$$\phi_x{}^* = \partial w/\partial x \text{ and } \phi_y{}^* = \partial w/\partial y.$$

6.2. Results for cylindrical arches and shells

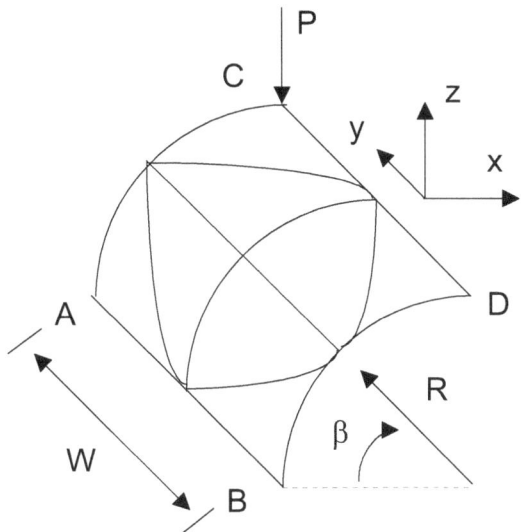

Figure 6.3. Mesh for section of a cylinder.

First the QBTP + DFT2 element is applied to analysis of half a semicircular three-pin arch, using automatically generated 'two-way' meshes (with crossed diagonals) three nodes wide, as illustrated in Figure 6.3. The boundary conditions and other data are

$$E = 10^6, \ v = 0, \ R = 10, \ W = 2, \ t = 1$$
$$u,v,w,\phi_z = 0 \text{ at base}, \quad u,v,\phi_z = 0 \text{ at crown} \tag{6.13}$$

The results for a mesh with 39 nodes (48 elements) are shown in Table 6.1 and compared to those when the DFT1 element of Section 5.2 is used. The DFT2 element gives significantly improved results.

The DFT2 stress results are average nodal values, averaged across the arch width. The DFT1 element gives poor nodal stresses for this problem, however, so that element centroidal stresses are used and these are averaged across the arch width for the four element centroids closest to the sampling location β.

As might be expected, both formulations also give satisfactory results for other cases, such as those of 2-pin and clamped arches, again the DFT2 formulation giving better results.

Table 6.1.
Results for 3 pin arch with DFT1/DFT2 elements (+ QBTP plate element).

Result	DFT1	DFT2	Theory
$w_C \times 10^{-4}$			
9 nodes	2.443	2.508	
15 nodes	3.791	3.823	
27 nodes	4.189	4.196	
39 nodes	4.190	4.263	4.248
39 nodes:			
m at β = 22.5 deg.	-1.534	-1.532	-1.533
m at β = 45 deg.	-2.040	-2.068	-2.071
m at β = 67.5 deg.	-1.535	-1.530	-1.533
T at β = 0	0.707	0.655	0.653
T at β = 45 deg.	0.767	0.703	0.707
T at β = 90 deg.	0.708	0.654	0.653

Next, using the same type of mesh shown in Figure 6.3, an octant of a cylindrical shell subject to an internal pressure $p = 0.5$ is analysed. For this the data and minimal boundary conditions are

$E = 1$, $v = 0$, $R = 2$, $W = 2$
v, w, ϕ_y, $\phi_z = 0$ at 'waist', u, v, ϕ_y, $\phi_z = 0$ at crown (6.14)

The nodal loads are exactly calculated as

$$q_x = - pRW(cos\delta\beta)/2, \quad q_y = pRW(sin\delta\beta)/2 \qquad (6.15)$$

where $\delta\beta$ is the angle of arc assigned to a node (that is, $\pi/2n$ when there are n rows of elements), this value being halved at edge nodes and halved again at corner nodes.

The expected radial displacement and circumferential stress are

$$\delta_r = pR^2/Et = 2, \quad \sigma_c = pR/t = 1 \qquad (6.16)$$

and the QBTP + DFT2 flat shell element gives similar solutions to the QBTP + DFT1 element, and these converge to the expected values.

With mesh refinement, however, stress accumulates at the central nodes (between AC and BD in Figure 6.3) and the remedy is to set u, ϕ_x, ϕ_z = 0 along AC and BD to model a section of a long cylinder. The results are then very close to the exact solutions, e.g. for 9 nodes δ_r = 2.0523, σ_c = 1.0262 and for 39 nodes δ_r = 2.0014, σ_c = 1.0071. Then for this special problem t = 0.1 → 0.0001 gives $\delta_r \times 10 \to 10,000$, as expected, without difficulty.

A more testing problem is that of a thin pinched cylinder and for this the problem considered by Ashwell and Sabir (1972) is considered. The boundary conditions are as in Equations 6.14 with u and ϕ_x = 0 also set along AC to model an octant of the cylinder and the other data is

E = 10.5 x 10^6, v = 0.3125, R = 4.953, W = 5.175, t = 0.01518
and P = 0.1 (1/4 applied) at C   (6.17)

Table 6.2. Results for deflection under load on thin pinched cylinder.

Nodes	DFT1 (+ QBTP)	DFT2 (+ QBTP)
9 (3 X 3)	0.021661	0.021661
15 (5 X 3)	0.024152	0.024126
27 (9 x 3)	0.024434	0.024471
39 (13 x 3)	0.024409	0.024472
Theory		0.02439

The results for the displacement under the load are shown in Table 6.2 for the formulations with the DFT1 and new DFT2 elements (combined with the QBTP element) and both formulations appear to perform satisfactorily (especially considering that only three nodes across the width are used), the results with DFT2 seemingly converged and those with DFT1 appear to be converging to a similar result.

6.3. Results for spherical shells

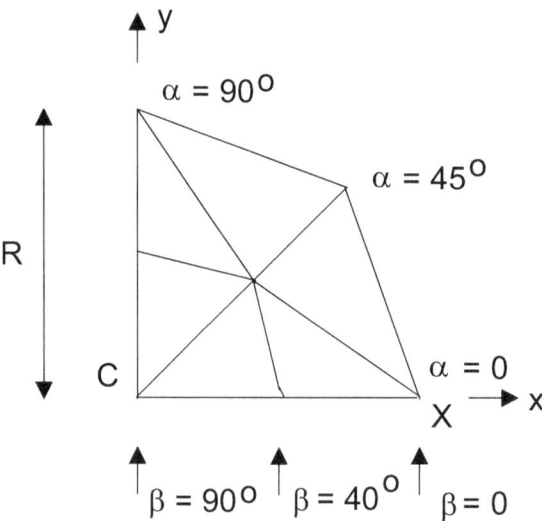

Figure 6.4. Spherical coordinates for coarse mesh for octant of a sphere.

Figure 6.4 is a plan view of the seven node mesh used for an octant of a spherical shell. The nodal coordinate data is spherical in α and β from which the Cartesian coordinates are easily calculated as

$$z = R\sin\beta, \ R_p = R\cos\beta, \ x = R_p \cos\alpha, \ y = R_p \sin\alpha \qquad (6.18)$$

The meshes used had 10, 13, 16, 25 and 34 nodes, $\delta\alpha = 22.5$ deg. (after the first $\delta\beta$), the increments in β, beginning from the crown C, being

10 nodes: $\delta\beta = 2\times22.5 + 45$; 13 nodes: $\delta\beta = 3\times18 + 36$;
16 nodes: $\delta\beta = 4\times15 + 30$; 25 nodes: $\delta\beta = 6\times10 + 2\times15$
34 nodes: $\delta\beta = 10\times7.5 + 15$

here using large elements near the 'waist' of the sphere to reduce the tendency for comparatively steep elements which tend to reduce solution accuracy.

The problem of a pinched sphere is considered with data and boundary conditions

$E = 1$, $v = 0$, $R = 4$, $t = 0.5$ and 0.05, $P = 1$ at C
$u, v, \phi_x, \phi_y, \phi_z = 0$ at C, $v, \phi_x = 0$ along CX (curved)
$u, \phi_y = 0$ along CY (curved), $w, \phi_z = 0$ along XY (curved)

combining these 'line' boundary conditions at points X and Y.

The results are given in Table 6.3 compared to the analytical solutions which are given by (Timoshenko & Woinowsky-Krieger, 1959)

$$w = \sqrt{3(1 - v^2)} \, PR/4Et^2 \tag{6.19}$$

and note that $P = 4$ (or unity in the FEM analysis for an octant of the sphere).

The numerical results are excellent with no more than 34 nodes and variations of the normal displacements close to the crown (for $\beta = 65 - 90$ deg.) are also close to the analytical solution.

The stress solutions are generally reasonable but a much finer mesh is needed near the crown where the problem may be treated as that of a shallow spherical cap to obtain solutions of greater accuracy (and in line with the manner in which the classical analytical solution is obtained).

Table 6.3. Deflections under load on a pinched sphere.

Nodes (mesh)	t = 0.5		t = 0.05	
	DFT1	DFT2	DFT1	DFT2
10 (2 x 3)	20.392	28.512	1166.3	1101.8
13 (2 x 4)	23.827	28.160	1632.9	1475.7
16 (2 x 5)	25.782	28.098	1945.6	1827.2
25 (2 x 8)	27.006	27.517	2549.7	2554.7
34 (2 x 11)	27.663	27.703	2766.3	2770.3
Theory		27.713		2771.3

Tables 6.4 and 6.5 show the results obtained by Olson and Bearden (1979) for a pinched sphere with $E = 1$, $v = 0.3$, $t = 1$, $R = 50$ and $P = -4$ (or $P = -1$ for an octant in the FEM model) and those obtained for this problem with the QBTP + DFT2 formulation.

Table 6.4. Olson and Bearden's results for a pinched sphere.

Mesh	Deflection at pole	Stresses at pole	Deflection at equator
2 x 2 (7 nodes)	24.42	-0.048/+0.1696	-1.164
4 x 4 (21 nodes)	32.14	0.2648/0.3784	-0.896
8 x 8 (73 nodes)	80.66	0.8440/0.9432	-0.796
10 x 10 (111 nodes)	87.46	0.9232/0.9848	-0.796
Theory (deep)	84.80	0.8248	-0.828

In Table 6.4 the deflections at the pole are reasonable though overshooting the solution. The stresses (circumferential and meridional) are poor, being slow to even approach the expected equality but the deflections at the equator are satisfactory.

Table 6.5. Present results (with QBTP + DFT2) for pinched sphere.

Mesh	Deflection at pole	Stresses at pole	Deflection at equator
2 x 3 (10 nodes)	52.10	0.3235	-0.954
2 x 4 (13 nodes)	66.00	0.3622	-1.332
2 x 5 (16 nodes)	75.41	0.3982	-1.735
2 x 8 (25 nodes)	80.22	0.4919	-1.423
2 x 11 (34 nodes)	78.08	0.5784	-1.337
4 x 4 (21 nodes)	50.12	0.3383/0.2789	-0.938
6 x 6 (43 nodes)	73.72	0.5818/4904	-0.783
Theory (deep)	84.80	0.8248	-0.828

In Table 6.5 two meshes are added to those used in Table 6.3, both using equal increments in β throughout. The deflections at the pole are reasonable, though an aberration in convergence occurs in the 25 node case, this being the only mesh in this group of 5 with two larger increments in β at the equator. The final two meshes show slower convergence, not being designed to give a good result for the pole deflection as the first five meshes were.

The stresses at the pole are equal, as required and are apparently converging satisfactorily though slowly. For the final two meshes the average nodal stresses are low and centroidal values are given and on a 'nodes used basis' are of comparable accuracy to those of Table 6.4. For the deflections at the equator, on the other hand, the first five meshes give poor results, whilst the final two meshes give satisfactory results, as expected.

Overall the results illustrate the difficulties and care needed in modeling such problems. Generally, however, the new element gives satisfactory results with remarkably few nodes and, in this respect at least, performs better than the very complicated Olson and Bearden solution.

As does the DFT2 formulation, the DFT1 formulation (+ QBTP) also gives satisfactory solutions to the problem of a pressurized sphere, but here the main accuracy problem is that of calculation of nodal loads. Overall, however, both elements perform satisfactorily over a range of shell problems but the DFT1 element does poorly on constant stress patch tests. The DFT2 element, however, passes these exactly, also performing slightly better on shell problems than the original element. As derivation of both the QBTP and DFT2 elements is remarkably simple, it is expected that their combined use as a facet shell element will be popular.

6.4. Program for the QBTP + DFT2 element

The VB5/6 coding for the QBTP+DFT2 element follows. Because the program was originally written in MegaBasic which automatically assigns upper/lower case according to context, there is some loss of consistency in appearance of upper and lower case.

The declarative statement DefDbl A-H, O-Z: DefInt I-N at the start of each subroutine is for double precision calculations and integer 'indexing'.

The first subroutine reads and calculates data:

(i) Line 110 sets degrees of freedom (NDF=6) per node and number of nodes per element (NCN=3).

(ii) Lines 120-140 give the local areal coordinates of the element numerical integration points.

(iii) Line 150 reads the problem size where NP = # nodes, NE = # elements, NB = number of boundary conditions points, NT = number of element property sets, NBW = half band width = NDF(max. node # diff. in any element + 1).

(iv) Line 160 reads the element property sets. (only one used here).

(v) Lines 170 to 320 generate the nodal coordinates and element node numbers for a regular cylindrical mesh. Here NX, NY = # of rows of nodes in each direction and XLIM, YLIM = $\pi R/2$, W in Figure 6.3 and if line 230 is omitted (by inserting REM) a planar mesh is generated. For spherical shells the spherical nodal coordinates are read as data and the Cartesian coordinates calculated from these.

(vi) Lines 332-355 calculate approximate loads for an internal pressure loading but are not required here.

(vii) Line 360 reads the nodal loads parallel to the Cartesian axes.

(viii) Line 380 calculates the *nodal valences*, the number of elements impinging at each node, for later use in calculating average nodal stresses.

```
[Attribute VB_Name = "Module1" - VB heading - not BASIC code]
DefDbl A-H, O-Z: DefInt I-N
Public NP, NE, NB, NT, NBW, NCN, NDF
Public NBC(50), Q(600), CI(6, 2), WF(6), FNE(100)
Public IMAT(100), CORD(100, 3), PROP(20, 5), NOP(100, 6)
Public op As Object: Public NFIX(50) As Long
Sub main()
Open "\newvb\cyldata.txt" For Input As #1
Open "\newvb\emats" For Output As #8
Open "\newvb\elstr" For Output As #9
Dim R(6): Set op = Form1: stt = Timer
110 NDF = 6: NCN = 3: EBC = 1: TOL = 0.0001
120 For I = 1 To 3: CI(I, 1) = 1 / 6: CI(I, 2) = 1 / 6: WF(I) = 1 / 6
CI(I + 3, 1) =0: CI(I + 3, 2) =0
130 WF(I + 3) = 0: Next: Rem Numerical integration data
140 CI(1, 1) = 4 / 6: CI(2, 2) = 4 / 6: CI(4, 1) = 1: CI(5, 2) = 1
150 Input #1, NP, NE, NB, NT, NBW: NBW = NBW - NDF
160 For N = 1 To NT: For I = 1 To 5: Input #1, PROP(N, I): Next: Next
170 Input #1, NX, NY, XLIM, YLIM
Rem NX, XLIM = # nodes & domain length in X direction
180 NEX = NX - 1: NEY = NY - 1: RNX = NEX: RNY = NEY
190 DX = XLIM / RNX: DY = YLIM / RNY: RS = XLIM * XLIM: DA = 2 * Atn(1) / (NX - 1)
200 For I = 1 To NX: Rem ANG=ANG+DA:If I=2 or I=3 then ANG=ANG+DA/2
205 For J = 1 To NY
210 RNDX = I - 1: RNDY = J - 1: NN = NY * (I - 1) + J
220 CORD(NN, 1) = RNDX * DX: CORD(NN, 2) = RNDY * DY
225 ANG = DA * (I - 1)
230 CORD(NN, 1) = XLIM * (1 - Cos(ANG)): CORD(NN, 3) = XLIM * Sin(ANG)
235 Next: Next
240 For I = 1 To NEX: For J = 1 To NEY: Rem Element nodes numbers are set up
250 NI = (I - 1) * NY + J: NJ = NY * I + J: FL = 0: If J > 1 Then EBC = EBC + 1
255 If EBC = 2 * Int(EBC / 2) Then FL = 1
260 NS = NEY * (I - 1) + J: NN = 2 * NS - 1
270 NOP(NN, 1) =NI: NOP(NN, 2) =NJ: NOP(NN, 3) =NI+1
If FL =1 Then NOP(NN, 3) =NJ+1
290 IMAT(NN) = 1: NN = 2 * NS
300 NOP(NN, 1) = NI + 1: NOP(NN, 2) = NJ: NOP(NN, 3) = NJ + 1
310 If FL <> 1 Then GoTo 320
315 NOP(NN, 1) = NI: NOP(NN, 2) = NJ + 1: NOP(NN, 3) = NI + 1
320 IMAT(NN) = 1: Next: Next: Rem Only one property set used
325 For I = 1 To NB: Input #1, NBC(I), NFIX(I): Next
330 GoTo 360
332 ANG = 2 * Atn(1) / (NX - 1): PL = XLIM * ANG / YLIM
334 For N = 1 To NP: CX = -(XLIM - CORD(N, 1)) / XLIM: CZ = CORD(N, 3) / XLIM
336 IC = (N - 1) * NDF + 1: NRL = PL: FX = 0: FY = 0
338 If CORD(N, 1) < TOL Or Abs(CORD(N, 1) - XLIM) < TOL Then FX = 1
340 If CORD(N, 2) < TOL Or Abs(CORD(N, 2) - YLIM) < TOL Then FY = 1
345 If FX = 1 Then NRL = NRL / 2
350 If FY = 1 Then NRL = NRL / 2
```

```
355 Q(IC) = Q(IC) + NRL * CX: Q(IC + 2) = Q(IC + 2) + NRL * CZ: Next
360 Input #1, NQ, R(1), R(2), R(3): Rem Read point loads
365 If NQ = 0 Then GoTo 380
370 For K = 1 To 3: IC = (NQ - 1) * NDF + K: Q(IC) = Q(IC) + R(K): Next
375 GoTo 360
380 For N = 1 To NE: For II = 1 To NCN: NN = NOP(N, II): FNE(NN) = FNE(NN) + 1
385 Next: Next: Rem calculate 'nodal valence'
390 For N = 1 To NE
400 Call newel(N): Next
410 Close: Open "\newvb\emats" For Input As #8
420 Open "\newvb\elstr" For Input As #9
430 Call fssolve
440 Call fsstress
450 op.Print "Time = ", (Timer - stt)
End Sub
```

The second subroutine is the element routine for the QBTP + DFT2 facet element. The main arrays are

ST, FT	the final element stiffness and force matrices (ESM and EFM)
SF, CF	the flexural ESM and EFM
SE, CE	the extensional ESM and EFM
S, C	the kernel ESM and EFM before basis transformation
T	the basis transformation matrix
D	the constitutive or modulus matrix
XY	the global element nodal coordinates
XYL	the local element nodal coordinates (in its plane)
TT	the coordinate transformation matrix of Equation 6.5
CT	the local direction cosines of the element sides
CI	the integration point coordinates
V	the interpolation matrix for the areal coordinate derivatives
Z	transforms from areal coordinate derivatives to Cartesian derivatives
G	the matrix H of Equation 5.24
PC	interpolation for the twist constraint of Equation 4.40 - not used here
PM	the flexural ESM when the latter is used

Then element data (for element N) is gathered in lines 50 - 90. The coordinate transformation is established in lines 100 - 138 and local coordinates and geometrical parameters calculated in lines 140-190.

The common (LST) kernel stiffness matrix for both flexure and extension is calculated by numerical integration in lines 240 - 440.

The basis transformation matrix for flexure (Equation 4.33) is formed in lines 450 -600 and used in lines 610 - 660. The basis transformation matrix for extension (Equation 5.26) is formed in lines 750 - 785 and used in lines 830 - 855. Then the ESMs for flexure and extension are combined to form the ESM ST( ) for the facet element in lines 860 - 875.

The basic 3×3 coordinate transformation matrix TT for coordinates and translational freedoms is used to form an 18 x 18 transformation for the whole element. Here a sign change is used to form the 3 x 3 matrix RT for rotational freedoms (following RHS rule) in line 138. Then matrices TT and RT are appropriately deployed to form the element coordinate transformation matrix TG in lines 880 - 895 and the latter used for form the final ESM and element force matrix EFM for the facet element in lines 897 - 980, these being written to storage in lines 1200 - 1230.

```
[Attribute VB_Name = "Module2"]
DefDbl A-H, O-Z: DefInt I-N
Sub newel(N)
10 Dim S(12, 12), C(9, 12), D(3, 3), B(3, 12), XY(3, 3), CT(3, 2), TT(3, 3), XYL(3, 3)
20 Dim TEMP(18, 18), T(12, 9), V(3, 6), Z(2, 3), SF(9, 9), CF(9, 9), G(6, 9), RT(3, 3)
30 Dim ST(18, 18), FT(18, 18), SE(12, 12), CE(9, 9), TG(18, 18), QE(3), TL(3, 3)
40 Dim PC(12), PM(12)
50 L = IMAT(N): E = PROP(L, 1): PR = PROP(L, 2): Rem Collect  properties for matrix D
60 DENS = PROP(L, 3): TH = PROP(L, 4): UDL = PROP(L, 5)
70 For M = 1 To 3: K = NOP(N, M): For I = 1 To 3: XY(I, M) = CORD(K, I): Next: Next
80 D(1, 1) = E * TH / (1 - PR * PR): D(2, 2) = D(1, 1)
D(1, 2) = PR * D(1, 1): D(2, 1) = D(1, 2)
90 D(3, 3) = 0.5 * (1 - PR) * D(1, 1): D(1, 3) = 0: D(2, 3) = 0: D(3, 1) = 0: D(3, 2) = 0
95 X21 = XY(1, 2) - XY(1, 1): X32 = XY(1, 3) - XY(1, 2): X13 = XY(1, 1) - XY(1, 3)
100 Y21 = XY(2, 2) - XY(2, 1): Y32 = XY(2, 3) - XY(2, 2): Y13 = XY(2, 1) - XY(2, 3)
110 A = X21 * Y32 - X32 * Y21: A = Abs(A)
115 TX = -(Y32 * XY(3, 1) + Y13 * XY(3, 2) + Y21 * XY(3, 3)) / A
120 CX = Sqr(1 / (1 + TX * TX)): SX = Sqr(1 - CX * CX): If TX < 0 Then SX = -SX
125 TY = (X32 * XY(3, 1) + X13 * XY(3, 2) + X21 * XY(3, 3)) / A
130 CY = Sqr(1 / (1 + TY * TY)): SY = Sqr(1 - CY * CY): If TY < 0 Then SY = -SY
132 TT(1, 1) = CX: TT(1, 2) = -SX * TY: TT(1, 3) = SX: Rem Fill coord. transf. matrix
134 TT(2, 1) = -SY * TX: TT(2, 2) = CY: TT(2, 3) = SY
136 TT(3, 1) = -CY * SX: TT(3, 2) = -CX * SY: TT(3, 3) = CX * CY
138 For J = 1 To 3: RT(3, J) = TT(3, J): RT(1, J) = TT(2, J): RT(2, J) = -TT(1, J): Next
140 For I = 1 To 3: For J = 1 To 3: XYL(I, J) = 0: For K = 1 To 3
145 XYL(I, J) = XYL(I, J) + TT(I, K) * XY(K, J): Next: Next: Next: Rem Calc. local coords
150 X21 = XYL(1, 2) - XYL(1, 1): Y21 = XYL(2, 2) - XYL(2, 1)
152 X32 = XYL(1, 3) - XYL(1, 2): Y32 = XYL(2, 3) - XYL(2, 2)
154 X13 = XYL(1, 1) - XYL(1, 3): Y13 = XYL(2, 1) - XYL(2, 3)
156 A = X21 * Y32 - X32 * Y21: A = Abs(A)
160 S21 = Sqr(X21 * X21 + Y21 * Y21): Rem element side lengths
170 S32 = Sqr(X32 * X32 + Y32 * Y32): S13 = Sqr(X13 * X13 + Y13 * Y13)
180 CT(1, 1) = X21 / S21: CT(1, 2) = Y21 / S21
185 CT(2, 1) = X32 / S32: CT(2, 2) = Y32 / S32: Rem Direction cosines of sides
190 CT(3, 1) = X13 / S13: CT(3, 2) = Y13 / S13
200 Rem QN=UDL*A/6:For I=1 to 3:NF=NOP(N,I)*NDF-3
210 Rem Q(NF)=Q(NF)+A*UDL/6:Next:Rem Add lumped loads to Q (plates only)
220 Z(1, 1) = -Y32 / A: Z(1, 2) = -Y13 / A: Z(1, 3) = -Y21 / A
Rem Transform from Cartesian
```

230 Z(2, 1) = X32 / A: Z(2, 2) = X13 / A: Z(2, 3) = X21 / A: Rem to local derivatives
240 For II = 1 To 6: Rem Commence integration loop ####################
250 F1 = 4 * CI(II, 1): F2 = 4 * CI(II, 2): F3 = 4 - F1 - F2
260 V(1, 1) = F1 - 1: V(1, 4) = F2: V(1, 6) = F3
270 V(2, 2) = F2 - 1: V(2, 4) = F1: V(2, 5) = F3: Rem Interp. for local derivatives
280 V(3, 3) = F3 - 1: V(3, 5) = F2: V(3, 6) = F1
290 For I = 1 To 2: For J = 1 To 6: TEMP(I, J) = 0: For K = 1 To 3
300 TEMP(I, J) = TEMP(I, J) + Z(I, K) * V(K, J): Next: Next: Next
310 For J = 1 To 6: JA = 2 * J - 1
315 PC(JA) = TEMP(2, J): PC(JA + 1) = -TEMP(1, J)
320 B(1, JA) = TEMP(1, J): B(2, JA + 1) = TEMP(2, J): Rem Fill strain interp. matrix
330 B(3, JA) = TEMP(2, J): B(3, JA + 1) = TEMP(1, J): Next
340 If II < 4 Then GoTo 380: Rem Form stress matrix for nodal stresses
350 For I = 1 To 3: IA = 3 * (II - 4) + I: For J = 1 To 12: C(IA, J) = 0: For K = 1 To 3
360 C(IA, J) = C(IA, J) + D(I, K) * B(K, J): Next: Next: Next
380 If WF(II) = 0 Then GoTo 440
390 For I = 1 To 3: For J = 1 To 12: TEMP(I, J) = 0: For K = 1 To 3
400 TEMP(I, J) = TEMP(I, J) + D(I, K) * B(K, J): Next: Next: Next
410 SMUL = WF(II) * A
420 For I = 1 To 12: For J = 1 To 12: For K = 1 To 3
425 S(I, J) = S(I, J) + B(K, I) * TEMP(K, J) * SMUL: Next: Next: Next
430 SMUL = 0 * SMUL * TH * TH * D(3, 3) / 12
435 For I = 1 To 12: For J = 1 To 12: Rem Add bending constraint matrix (not used)
437 PM(I, J) = S(I, J) + SMUL * PC(I) * PC(J): Next: Next
440 Next II: Rem End integration loop ############################
450 T(1, 2) = 1: T(2, 3) = 1: T(3, 5) = 1: T(4, 6) = 1: T(5, 8) = 1: T(6, 9) = 1
460 T(7, 1) = -1.5 * X21 / S21 ^ 2: T(7, 4) = -T(7, 1): Rem Fill basis transformation matrix
470 T(8, 1) = -1.5 * Y21 / S21 ^ 2: T(8, 4) = -T(8, 1)
480 T(7, 2) = 0.5 * CT(1, 2) ^ 2 - 0.25 * CT(1, 1) ^ 2: T(7, 5) = T(7, 2)
490 T(7, 3) = -0.75 * CT(1, 1) * CT(1, 2): T(7, 6) = T(7, 3): T(8, 2) = T(7, 3)
500 T(8, 5) = T(7, 3): T(8, 3) = 0.5 * CT(1, 1) ^ 2 - 0.25 * CT(1, 2) ^ 2: T(8, 6) = T(8, 3)
510 T(9, 4) = -1.5 * X32 / S32 ^ 2: T(9, 7) = -T(9, 4)
520 T(10, 4) = -1.5 * Y32 / S32 ^ 2: T(10, 7) = -T(10, 4)
530 T(9, 5) = 0.5 * CT(2, 2) ^ 2 - 0.25 * CT(2, 1) ^ 2: T(9, 8) = T(9, 5)
540 T(9, 6) = -0.75 * CT(2, 1) * CT(2, 2): T(9, 9) = T(9, 6): T(10, 5) = T(9, 6)
550 T(10, 8) = T(9, 6): T(10, 6) = 0.5 * CT(2, 1) ^ 2 - 0.25 * CT(2, 2) ^ 2
T(10, 9) = T(10, 6)
560 T(11, 1) = 1.5 * X13 / S13 ^ 2: T(11, 7) = -T(11, 1)
570 T(12, 1) = 1.5 * Y13 / S13 ^ 2: T(12, 7) = -T(12, 1)
580 T(11, 2) = 0.5 * CT(3, 2) ^ 2 - 0.25 * CT(3, 1) ^ 2: T(11, 8) = T(11, 2)
590 T(11, 3) = -0.75 * CT(3, 1) * CT(3, 2): T(11, 9) = T(11, 3): T(12, 2) = T(11, 3)
600 T(12, 8) = T(11, 3): T(12, 3) = 0.5 * CT(3, 1) ^ 2 - 0.25 * CT(3, 2) ^ 2
T(12, 9) = T(12, 3)
605 FMUL = TH * TH / 12
610 For I = 1 To 12: For J = 1 To 9: TEMP(I, J) = 0: For K = 1 To 12
620 TEMP(I, J) = TEMP(I, J) + PM(I, K) * T(K, J) * FMUL: Next: Next: Next
630 For I = 1 To 9: For J = 1 To 9: SF(I, J) = 0: For K = 1 To 12
640 SF(I, J) = SF(I, J) + T(K, I) * TEMP(K, J): Next: Next: Next: Rem Flexural ESM
650 For I = 1 To 9: For J = 1 To 9: CF(I, J) = 0: For K = 1 To 12
660 CF(I, J) = CF(I, J) + C(I, K) * T(K, J) * FMUL: Next: Next: Next

Rem Flexural stress matrix
750 For I = 1 To 6: For J = 1 To 9: G(I, J) = 0: Next: Next
751 Rem Form basis transformation matrix for plane stress
752 G(1, 1) = CT(1, 1) / 2: G(1, 2) = CT(1, 2) / 2: G(1, 4) = CT(1, 1) / 2: G(1, 5) = CT(1, 2) / 2
754 G(3, 4) = CT(2, 1) / 2: G(3, 5) = CT(2, 2) / 2: G(3, 7) = CT(2, 1) / 2: G(3, 8) = CT(2, 2) / 2
755 G(5, 1) = CT(3, 1) / 2: G(5, 2) = CT(3, 2) / 2: G(5, 7) = CT(3, 1) / 2: G(5, 8) = CT(3, 2) / 2
757 G(2, 1) = -CT(1, 2) / 2: G(2, 2) = CT(1, 1) / 2: G(2, 4) = -CT(1, 2) / 2: G(2, 5) = CT(1, 1) / 2
758 G(4, 4) = -CT(2, 2) / 2: G(4, 5) = CT(2, 1) / 2: G(4, 7) = -CT(2, 2) / 2: G(4, 8) = CT(2, 1) / 2
759 G(6, 1) = -CT(3, 2) / 2: G(6, 2) = CT(3, 1) / 2: G(6, 7) = -CT(3, 2) / 2: G(6, 8) = CT(3, 1) / 2
760 G(2, 3) = S21 / 8: G(2, 6) = -S21 / 8: G(4, 6) = S32 / 8: G(4, 9) = -S32 / 8
762 G(6, 3) = -S13 / 8: G(6, 9) = S13 / 8
765 For I = 1 To 6: For J = 1 To 6: T(I, J) = 0: Next: Next
770 T(1, 1) = CT(1, 1): T(1, 2) = -CT(1, 2): T(2, 1) = CT(1, 2): T(2, 2) = CT(1, 1)
772 T(3, 3) = CT(2, 1): T(3, 4) = -CT(2, 2): T(4, 3) = CT(2, 2): T(4, 4) = CT(2, 1)
775 T(5, 5) = CT(3, 1): T(5, 6) = -CT(3, 2): T(6, 5) = CT(3, 2): T(6, 6) = CT(3, 1)
776 For I = 1 To 6: For J = 1 To 9: TEMP(I, J) = 0: For K = 1 To 6
778 TEMP(I, J) = TEMP(I, J) + T(I, K) * G(K, J): Next: Next: Next
780 For I = 1 To 12: For J = 1 To 9: T(I, J) = 0: Next: Next
782 T(1, 1) = 1: T(2, 2) = 1: T(3, 4) = 1: T(4, 5) = 1: T(5, 7) = 1: T(6, 8) = 1
785 For I = 1 To 6: IA = I + 6: For J = 1 To 9: T(IA, J) = TEMP(I, J): Next: Next
830 For I = 1 To 12: For J = 1 To 9: TEMP(I, J) = 0: For K = 1 To 12
835 TEMP(I, J) = TEMP(I, J) + S(I, K) * T(K, J): Next: Next: Next
840 For I = 1 To 9: For J = 1 To 9: SE(I, J) = 0: For K = 1 To 12
845 SE(I, J) = SE(I, J) + T(K, I) * TEMP(K, J): Next: Next: Next: Rem extensional ESM
850 For I = 1 To 9: For J = 1 To 9: CE(I, J) = 0: For K = 1 To 12
855 CE(I, J) = CE(I, J) + C(I, K) * T(K, J): Next: Next: Next: Rem extensional stress matrix
860 For II = 1 To 3: For JJ = 1 To 3: For I = 1 To 3: IA = (II - 1) * 3 + I
865 NR = (II - 1) * 6 + I: NI = NR + 2: If I = 3 Then NR = NR + 3
868 For J = 1 To 3: NC = (JJ - 1) * 6 + J: NJ = NC + 2: If J = 3 Then NC = NC + 3
870 JA = (JJ - 1) * 3 + J: ST(NR, NC) = SE(IA, JA)
875 ST(NI, NJ) = SF(IA, JA): Next: Next: Next: Next
880 For NN = 1 To 6: For I = 1 To 3: For J = 1 To 3
885 NR = (NN - 1) * 3 + I: NC = (NN - 1) * 3 + J
890 MF = 1: If NN = 2 * Int(NN / 2) Then MF = 0
895 TG(NR, NC) = MF * TT(I, J) + (1 - MF) * RT(I, J): Next: Next: Next
897 For I = 1 To 18: For J = 1 To 18: TEMP(I, J) = 0: For K = 1 To 18
900 TEMP(I, J) = TEMP(I, J) + ST(I, K) * TG(K, J): Next: Next: Next
910 For I = 1 To 18: For J = 1 To 18: ST(I, J) = 0: For K = 1 To 18
920 ST(I, J) = ST(I, J) + TG(K, I) * TEMP(K, J): Next: Next: Next
925 For I = 1 To 18: For J = 1 To 18: TEMP(I, J) = 0: Next: Next
927 For K = 1 To 3
930 For JJ = 1 To 3: For IB = 1 To 3: I = 6 * K - 6 + IB: IA = I + 3: IC = 3 * K - 3 + IB
940 For J = 1 To 3: NC = (JJ - 1) * 6 + J: NJ = NC + 2: If J = 3 Then NC = NC + 3
950 JA = (JJ - 1) * 3 + J: TEMP(I, NC) = CE(IC, JA)
960 TEMP(IA, NJ) = CF(IC, JA): Next: Next: Next
965 Next K
970 For I = 1 To 18: For J = 1 To 18: FT(I, J) = 0: For K = 1 To 18
980 FT(I, J) = FT(I, J) + TEMP(I, K) * TG(K, J): Next: Next: Next
1000 Rem PF=ST(1,1)/10000:ST(6,6)=PF:ST(12,12)=PF:ST(18,18)=PF

```
1200 For I = 1 To 18: For J = 1 To 18
1210 Write #8, ST(I, J): Next: Next: Rem File the ESMs
1220 For I = 1 To 18: For J = 1 To 18
1230 Write #9, FT(I, J): Next: Next: Rem File stress matrices
Debug.Print N
End Sub
```

The third subroutine reads (line 170), assembles the banded system equations (lines 180 - 240) and solves them using Gauss reduction (lines 280 - 550), storing the reduced equations for 'block solution' (Mohr, 1992). Equations with boundary conditions are flagged with NOB=1 in lines 300 - 360 for reversal of sign, enabling their later use to calculate boundary 'reactions' (not used here).

Thus the equations are stored in a block of size SIZ x LB (set in line 20). When the block is full IBUF (reduced) equations are written to file 'stifm' (lines 560 - 640) and then read back (lines 820 - 880) in reverse order during back substitution (lines 700 - 920).

```
[Attribute VB_Name = "Module3"]
DefDbl A-H, O-Y: DefInt I-N
Public DIS(6, 100): Private ZI, ZJ As Long
Sub fssolve()
Open "\newvb\stifm" For Random As #10 Len = 600
Open "\newvb\vbout" For Output As #7
10 Dim ESM(18, 18), SK(120, 60), SKP(60), T(60)
20 SIZ = 120: LB = 60: IBUF = SIZ - LB: NRW = 0: NTW = NBW + NDF
30 NLOAD = NP * NDF: NBN = 1: L = 0: N = 1
100 L = L + 1: If N > NE Then GoTo 280
Rem Commence loop for node by node forward reduction ###########
110 For M = 1 To 8: Rem Search for up to 8 eles introduced by this node
130 For I = 1 To NCN: If NOP(N, I) = L Then GoTo 170
150 Next: Rem Check if next ESM introduced by this node
160 GoTo 280
170 For I = 1 To 18: For J = 1 To 18: Input #8, ESM(I, J): Next: Next: Rem Read ESMs
180 For I = 1 To NCN: For J = 1 To NCN
190 For IL = 1 To NDF: IE = (I - 1) * NDF + IL: NR = (NOP(N, I) - 1) * NDF + IL
200 NRE = NR - NRW: Rem NRW = # of rows of SSM filed
210 For JL = 1 To NDF: JE = (J - 1) * NDF + JL: NC = (NOP(N, J) - 1) * NDF + JL
220 NCB = NC - NR + 1: If NR > NC Then GoTo 240
230 SK(NRE, NCB) = SK(NRE, NCB) + ESM(IE, JE): Rem Assembly of SSM
240 Next: Next: Next: Next
270 N = N + 1: Next M
280 NDIF = (NP - L + 1) * NDF: If NDIF > NBW Then LIM = NBW + NDF
300 ZJ = 0: If NBN = NB + 1 Then GoTo 320
305 If L <> NBC(NBN) Then GoTo 320
310 ZJ = NFIX(NBN): ZI = 100000: NBN = NBN + 1: Rem Check if node has b.c.'s
```

```
320 For ID = 1 To NDF
330 LIM = LIM - 1: IP = ID + NDF * (L - 1): IPE = IP - NRW: R = Q(IP): NOB = 0
340 If ZJ = 0 Then GoTo 380
345 If ZJ < ZI Then GoTo 360: Rem Check this freedom for b.c.
350 RS = -R: R = 0: NOB = 1: ZJ = ZJ - ZI
360 ZI = ZI / 10
380 If NOB = 1 Then GoTo 420: Rem NOB = boundary flag
390 XK = 1 / SK(IPE, 1): Q(IP) = XK * R:  GoTo 440
420 Q(IP) = RS + SK(IPE, 1) * R: XK = -1: R = -R: Rem Q(IP) = boundary 'reaction'
440 For J = 1 To LIM: JA = J + 1: SKP(J) = SK(IPE, JA): Next: Rem Store 'row multipliers'
450 NC = LIM + 1: Rem  XK = -1 for b.c. gives - boundary row to detect in backsub.
460 For J = 1 To NC: SK(IPE, J) = SK(IPE, J) * XK: Next: Rem Divide row by pivot
480 If (L + ID - NP - NDF) = 0 Then GoTo 660: Rem End test
490 For I = 1 To LIM: If SKP(I) = 0 Then GoTo 550
505 If NOB = 1 Then GoTo 530
510 NC = LIM - I + 1: NRE = IPE + I
515 For J = 1 To NC: JP = J + I
520 SK(NRE, J) = SK(NRE, J) - SK(IPE, JP) * SKP(I): Next: Rem Forward reduction
530 JP = I + 1: NR = IP + I: Q(NR) = Q(NR) - SK(IPE, JP) * R
550 Next I
560 If (IPE + NTW) < SIZ Then GoTo 640: Rem Test if stiffness block full
570 If (NLOAD - NRW) <= SIZ Then GoTo 640
575 NBLOCK = NBLOCK + 1
580 For I = 1 To IBUF: NREC = (NBLOCK - 1) * IBUF + I
590 For J = 1 To LB: T(J) = SK(I, J): Next
595 Put #10, NREC, T: Next: Rem File part of stiffness block
600 NFLAG = NBLOCK: NRW = NRW + IBUF: Rem NRW = # of rows of SSM filed
610 For I = 1 To LB: For J = 1 To LB: IA = I + IBUF
620 SK(I, J) = SK(IA, J): SK(IA, J) = 0: Next: Next: Rem Shift remaining rows up
640 Next ID
650 GoTo 100: Rem End node by node forward reduction loop ##############
660 NR =NDF * NP: NRE =NR - NRW: DIS(NDF, NP) =Q(NR)
Rem Last displacement known
680 Q(NR) = 0: I = NDF: L = NP: GoTo 780
700 L = L - 1: Rem Loop on nodes for back substitution
710 I = I - 1: Rem Loop on d.f./node for back substitution
720 NR = NDF * (L - 1) + I: NRE = NR - NRW: DIS(I, L) = Q(NR): Q(NR) = 0
740 If LIM < (NBW + NDF - 1) Then LIM = LIM + 1
750 For J = 1 To LIM: JA = J + 1
760 LJ = L + Int((J + I - 1) / NDF): K = I + J - (LJ - L) * NDF
770 DIS(I, L) = DIS(I, L) - SK(NRE, JA) * DIS(K, LJ): Next: Rem Backsub. calculation
780 If SK(NRE, 1) > 0 Then GoTo 800
790 Q(NR) = DIS(I, L): DIS(I, L) = 0: Rem Set suppressed displacements
800 If (NRE - NTW) > 0 Or NRW = 0 Then GoTo 890
820 For II = 1 To LB: IA = SIZ - II + 1: IB = LB - II + 1: For J = 1 To LB
840 SK(IA, J) = SK(IB, J): Next: Next
850 NRW = NRW - IBUF
855 For II = 1 To IBUF: Rem Read back filed part of SSM
860 NREC = (NBLOCK - 1) * IBUF + II: Get #10, NREC, T
870 For J = 1 To LB: SK(II, J) = T(J): Next: Next
```

```
880 NBLOCK = NBLOCK - 1
890 If (I + L - 2) = 0 Then GoTo 930: Rem End test
900 If I <> 1 Then GoTo 710: Rem End loop on freedoms/node
910 I = NDF + 1
920 GoTo 700: Rem End backsub. loop on nodes
930 op.Print "NODAL DISPLACEMENTS U,V,W": a$ = " #.####E+00;-#.####E+00"
940 For N = 1 To NP
950 op.Print N, Format(DIS(1, N), a$), Format(DIS(2, N), a$), Format(DIS(3, N), a$)
960 Write #7, N, DIS(1, N), DIS(2, N), DIS(3, N): Next
990 op.Print "# Blocks =", NFLAG
End Sub
```

The fourth and final subroutine reads the element force matrices (EFMs) from storage and calculates the element stresses by multiplying these by the displacement solutions for each element's nodes (line 130).

Finally nodal average stresses are calculated in line 170.

```
Attribute VB_Name = "Module4"
DefDbl A-H, O-Z: DefInt I-N
Sub fsstress()
10 Dim F(18), B(18, 18), R(18), FORCE(100, 6)
20 For N = 1 To NE
30 For I = 1 To 18: For J = 1 To 18
40 Input #9, B(I, J): Next: Next: Rem Read element stress matrix
50 For I = 1 To NCN: M = NOP(N, I)
60 If M = 0 Then GoTo 100
70 K = (I - 1) * NDF
80 For J = 1 To NDF: IJ = J + K
90 R(IJ) = DIS(J, M): Next J: Rem Collect element displacements
100 Next I
110 IA = K + NDF
120 For I = 1 To 18: F(I) = 0: For J = 1 To IA
130 F(I) = F(I) + B(I, J) * R(J): Rem Calculate element stresses
140 Next: Next
150 For II = 1 To 3: NI = NOP(N, II)
160 For I = 1 To 6: IA = 6 * (II - 1) + I
170 FORCE(NI, I) = FORCE(NI, I) + F(IA) / FNE(NI): Rem Average nodal stresses
190 Next: Next
200 Rem For I=1 to 6:FORCE(N,I)=F(I):Next
210 Next N
220 op.Print "NODAL AVERAGE DIRECT (X&Y) AND TWIST MOMENTS"
230 a$ = " #.####E+00;-#.####E+00"
240 For N = 1 To NP: op.Print N, Format(FORCE(N, 1), a$), Format(FORCE(N, 2), a$), _
Format(FORCE(N, 4), a$), Format(FORCE(N, 5), a$)
250 Write #7, N, FORCE(N, 1), FORCE(N, 2), FORCE(N, 4), FORCE(N, 5): Next
End Sub
```

The data file cyldata.txt called in the first subroutine is

```
9,8,7,1,30
10.5e6,0.3125,0,0.01548,0
3,3,4.953,5.175
1,011011
2,011011
3,011111
6,010100
7,110011
8,110011
9,110111
9,0,0,-0.025
0,0,0,0
```

this being the data for pinched cylinder problem of Equations 6.17 with 9 nodes and the solution for the displacement under the load should be that given in Table 6.2.

Here the first line gives the problem size, the second the element properties, the third the mesh and domain size. The following 8 lines give the boundary condition flags for which 1 = fixed and 0 = free for $u, v, w, \phi_x, \phi_y, \phi_z$ and the final lines specify point loads corresponding to the x,y,z axes and are terminated by a row of zeros, terminating reading of data.

6.5. Conclusions

The QBTP + DFT2 element is relatively simple, particularly because both component elements have the same kernel stiffness matrix (of the LST element), and should prove popular.

As this text concentrates on *natural strains*, as well as basis transformation, it uses the displacement method, but note that hybrid and mixed formulations have also been obtained which include the drilling freedom in facet shell elements (Saleeb et al, 1988; Allman, 1994; To & Liu, 1998).

6.6. References

Allman, A basic flat finite element for the analysis of general shells, *Int. J. Num. Meth. Engng* 37 (1994) 169-203.

Ashwell DG, Sabir AB. A new cylindrical shell finite element based in simple independent strain functions, *Int. J. Mech. Sci.* 14 (1972) 171.

Clough and Johnson, A finite element approximation for the analysis of thin shells, *Int. J. Solids & Structures*, 4 (1968) 43.

Mohr GA, *Analysis and Design of Plate and Shell Structures using Finite Elements*, PhD thesis, University of Cambridge 1976.

Mohr GA, *Finite Elements for Solids, Fluids, and Optimization, Oxford University Press*, OUP 1992.

Mohr GA, An improved facet shell element, *Int. J. Arts & Sciences,* vol. 1, no. 1, pp 19-26 (2001).

Mohr GA, A new facet shell element, *Int. J. Arts & Sciences,* vol. 1, no. 2, pp 36- 49 (2001).

Mohr GA, An accurate facet shell element, *Int. J. Arts & Sciences,* vol. 2, no. 1, pp 1-13 (2002).

Olsen M, Bearden TW, A simple flat triangular flat shell element revisited, Int. J. Numerical Methods Engng 14 (1979) 51.

Parekh CJ, *Finite Element Solution System*, PhD thesis, University of Wales, Swansea, 1969.

Saleeb, TY, Chang TY, Yinuengyong, A mixed formulation C^0 linear triangular plate/shell element, the role of edge shear constraints, *Int. J. Numerical Methods in Engineering* 16, 1988, 1101.

Timoshenko SP, Woinowsky-Krieger WK, *Theory of Plates and Shells*, 2nd edn, McGraw-Hill, New York 1959.

To WS, Liu ML, A further study of hybrid strain-based three-node flat triangular shell elements, *Finite Elements in Analysis & Design* 31, 1998, 135.

Zienkiewicz OC, *The Finite Element Method*, 3rd edn, McGraw-Hill, London 1977.

6. Facet Shell Elements

Chapter 7

THICK PLATE ELEMENTS

7.1. Introduction

Several thick plate finite elements have been developed (Griemann & Lynn, 1970; Pryor et al; 1970, Rao et al; 1974, Pugh et al, 1978).

In thick plate elements the deformations arising from transverse shearing are taken into account. Then the generalized strains are

$$\chi_x = \partial\phi_x/\partial x, \quad \chi_y = \partial\phi_y/\partial y, \quad \chi_{xy} = \partial\phi_x/\partial y + \partial\phi_y/\partial x$$

$$(7.1a)$$

$$\gamma_x = \partial w/\partial x - \phi_x, \quad \gamma_y = \partial w/\partial x - \phi_y \qquad (7.1b)$$

where w is the total deflection, ϕ_x and ϕ_y are the slopes caused by flexing of the plate and γ_x and γ_y are the additional slopes arising from deformation of the plate caused by transverse shearing. From simple energy arguments (Mohr, 1992) it also follows that γ_x and γ_y are also the average transverse shearing strains.

The constitutive equations for thick plates are then

$$
\begin{Bmatrix} m_x \\ m_y \\ m_{xy} \\ Q_x \\ Q_y \end{Bmatrix} = Et/12(1-v^2)
\begin{bmatrix}
t^2 & vt^2 & 0 & 0 & 0 \\
vt^2 & t^2 & 0 & 0 & 0 \\
0 & 0 & \frac{1}{2}(1-v)t^2 & 0 & 0 \\
0 & 0 & 0 & 5(1-v) & 0 \\
0 & 0 & 0 & 0 & 5(1-v)
\end{bmatrix}
\begin{Bmatrix} \chi_x \\ \chi_y \\ \chi_{xy} \\ \gamma_x \\ \gamma_y \end{Bmatrix} = D\{\varepsilon\}
$$

$$(7.2)$$

Equations 7.1 and 7.2 omit only one of the three dimensional strains, that of transverse compression. This leaves them without the third 'coupling' shear strain $\varepsilon_{xy} = \partial u/\partial y + \partial v/\partial x$ as in plane stress elements.

This sometimes leads to numerical difficulties when triangular thick plate elements are formulated and in the following section an 18 freedom triangular thick plate element is developed using the natural strain approach for both curvatures and shear strains (Argyris & Scharpf, 1971).

Then by applying penalty factors to enforce the Kirchhoff constraints

$$\partial w/\partial x - \phi_x = 0, \quad \partial w/\partial x - \phi_y = 0 \tag{7.3}$$

and thus suppress the shear strains, the element can also be used to model thin plates.

7.2. Triangular thick plate element

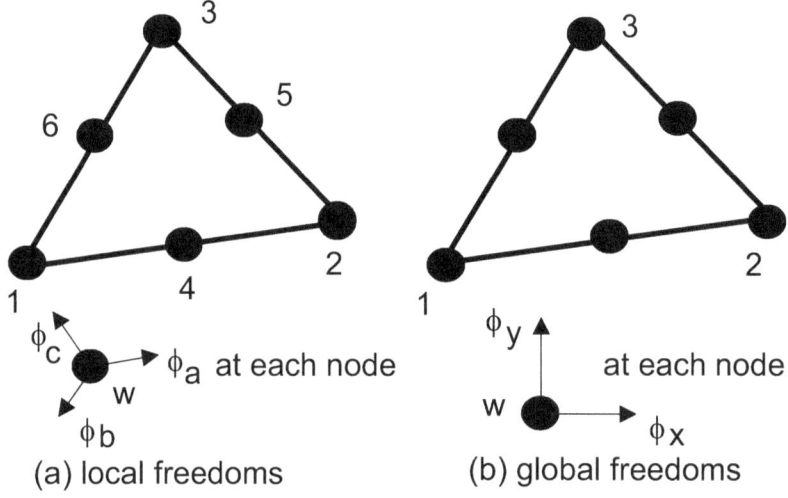

Figure 7.1. Thick plate element.

Figure 7.1 shows a triangular thick plate element with 18 freedoms, w, ϕ_x, ϕ_y at six nodes. Transforming the Cartesian global slopes to natural slopes parallel to the sides

$$\phi_a = c_{ax}\phi_x + c_{ay}\phi_y, \quad \phi_b = c_{bx}\phi_x + c_{by}\phi_y, \quad \phi_c = c_{cx}\phi_x + c_{cy}\phi_y \tag{7.4}$$

where

$$c_{ax} = x_{21}/L_a, \; c_{ay} = y_{21}/L_b, \; c_{bx} = x_{32}/L_b, \; c_{by} = y_{32}/L_b, c_{cx} = x_{13}/L_c, \; c_{cy} = y_{13}/L_c,$$

The interpolation for the displacement and natural slopes is that of the linear strain triangle (LST)

$$w = \Sigma f_i w_i, \quad \phi_a = \Sigma f_i \phi_{ai}, \quad \phi_b = \Sigma f_i \phi_{bi}, \quad \phi_c = \Sigma f_i \phi_{ci} \quad i = 1 \to 6$$
(7.5)

where

$$f_1 = 2L_1^2 - 1, \quad f_2 = 2L_2^2 - 1, \quad f_3 = 2L_3^2 - 1, \quad f_4 = 4L_1 L_2, \quad f_5 = 4L_2 L_3, \quad f_6 = 4L_3 L_1$$
(7.6)

Then the natural curvatures and shear strains are given by

$$\chi_a = \partial\phi_a/\partial a = [\partial\phi_a/\partial L_2 - \partial\phi_a/\partial L_1]/L_a$$

$$\chi_b = \partial\phi_b/\partial b = [\partial\phi_b/\partial L_3 - \partial\phi_b/\partial L_2]/L_b$$

$$\chi_c = \partial\phi_c/\partial c = [\partial\phi_c/\partial L_1 - \partial\phi_c/\partial L_3]/L_c$$

$$\gamma_a = \partial w/\partial a - \phi_a = [\partial w/\partial L_2 - \partial w/\partial L_1]/L_a - \phi_a$$

$$\gamma_b = \partial w/\partial b - \phi_b = [\partial w/\partial L_3 - \partial w/\partial L_2]/L_b - \phi_b$$

$$\gamma_c = \partial w/\partial c - \phi_c = [\partial w/\partial L_1 - \partial w/\partial L_3]/L_c - \phi_c$$
(7.7)

Combining Equations 7.4 - 7.7 the 6×18 interpolation matrix B for the natural generalized strains is given by

$$\{\varepsilon\} = \begin{Bmatrix} \chi_a \\ \chi_b \\ \chi_c \\ \gamma_a \\ \gamma_b \\ \gamma_c \end{Bmatrix} = B\{d\} = \begin{bmatrix} B_b \\ B_s \end{bmatrix}\{d\} = \begin{bmatrix} w\ columns & \phi_x\ columns & \phi_y\ columns \\ 0 & c_{ax}\{f_a\}^t & c_{ay}\{f_a\}^t \\ 0 & c_{bx}\{f_b\}^t & c_{by}\{f_b\}^t \\ 0 & c_{cx}\{f_c\}^t & c_{cy}\{f_c\}^t \\ \{f_a\}^t & -c_{ax}\{f\}^t & -c_{ay}\{f\}^t \\ \{f_b\}^t & -c_{bx}\{f\}^t & -c_{by}\{f\}t \\ \{f_c\}^t & -c_{cx}\{f\}^t & -c_{cy}\{f\}^t \end{bmatrix} \{d\}$$
(7.8)

where

$$\{f_a\} = \partial\{f\}/\partial a = \{1-4L_1, \ 4L_2-1, \ 0, \ 4L_1-4L_2, \ 4L_3, \ -4L_3\}/L_a$$

$$\{f_b\} = \partial\{f\}/\partial b = \{0, \ 1-4L_2, \ 4L_3-1, \ -4L_1, \ 4L_2-4L_3, \ 4L_1\}/L_b$$

$$\{f_c\} = \partial\{f\}/\partial c = \{4L_1-1, \ 0, \ 1-4L_3, \ 4L_2, -4L_2, \ 4L_3-4L_1\}/L_c$$
(7.9)

Transformation of the natural curvatures given by B_b to Cartesian values is accomplished using matrix C_N^{-1} of 3.47 and the transformation for the shears is obtained by writing (Argyris and Scharpf, 1971):

$$\gamma_a = c_{ax}\gamma_x + c_{ay}\gamma_y, \quad \gamma_b = c_{bx}\gamma_x + c_{by}\gamma_y \tag{7.10}$$

or $\gamma_b/c_{bx} = \gamma_x + c_{by}\gamma_y/c_{bx}, \quad \gamma_a/c_{ax} = \gamma_x + c_{ay}\gamma_y/c_{ax}$ $\qquad$ (7.11)

and eliminating γ_x from Equations 7.11

$$\gamma_y(c_{ax}c_{by} - c_{bx}c_{ay}) = \gamma_b c_{ax} - \gamma_a c_{bx} \tag{7.12}$$

The area of the triangular element is given by

$$2\Delta = x_{21}y_{32} - x_{32}y_{21} = L_a L_b(c_{ax}c_{by} - c_{bx}c_{ay}) \tag{7.13}$$

so that Equation 7.12 can be written

$$\gamma_y = L_a L_b(-c_{bx}\gamma_a + c_{ax}\gamma_b)/2\Delta \tag{7.14a}$$

In like fashion Equations 7.10 can be solved for γ_x

$$\gamma_x = L_a L_b(c_{by}\gamma_a - c_{ay}\gamma_b)/2\Delta \tag{7.14b}$$

Using cyclic progression two further permutations of Equations 7.14 are obtained

$$\gamma_x = L_b L_c(c_{cy}\gamma_b - c_{by}\gamma_c)/2\Delta, \quad \gamma_y = L_b L_c(-c_{cx}\gamma_b + c_{bx}\gamma_c)/2\Delta \tag{7.15}$$

$$\gamma_x = L_c L_b(c_{ay}\gamma_c - c_{cy}\gamma_a)/2\Delta, \quad \gamma_y = L_c L_a(-c_{ax}\gamma_c + c_{cx}\gamma_a)/2\Delta \tag{7.16}$$

Averaging Equations 7.14 - 716 yields the required transformation

$$\left\{\begin{array}{c}\gamma_x \\ \gamma_y\end{array}\right\} = \left(\frac{1}{6\Delta}\right)\left[\begin{array}{ccc} c_{by}L_a L_b - c_{cy}L_c L_a & c_{cy}L_b L_c - c_{ay}L_a L_b & c_{ay}L_c L_a - c_{by}L_b L_c \\ c_{cx}L_c L_a - c_{bx}L_a L_b & c_{ax}L_a L_b - c_{cx}L_b L_c & c_{bx}L_b L_c - c_{ax}L_c L_a \end{array}\right]\left\{\begin{array}{c}\gamma_a \\ \gamma_b \\ \gamma_c\end{array}\right\} = T_s\{\gamma_N\}$$

$$\tag{7.17}$$

Thus the transformed strain-displacement matrix for both bending and shear is given as

$$B = \begin{bmatrix} B_b{}^* \\ B_s{}^* \end{bmatrix} = \begin{bmatrix} C_N^{-1} B_b \\ T_s B_s \end{bmatrix} \tag{7.18}$$

and, including a penalty factor β the final element stiffness matrix is given by three point numerical integration

$$k = \Sigma \, [B_b^t{}^* \, D_b \, B_b{}^* + \beta B_s^t{}^* D_s B_s{}^*] \, (\varpi_i \Delta) \tag{7.19}$$

using integration at three internal points given by the areal coordinates

$$(L_1, L_2, L_3) = (4/6, 1/6, 1/6), \, (1/6, 4/6, 1/6), \, (1/6, 1/6.4/6) \tag{7.20}$$

with weights $\varpi_i = 1/3$ at each. In Equation 7.19 D_b is the upper left 3×3 part of matrix D of Equation 7.2 and D_s is the lower right 2×2 part of D.

This integration is derived by Mohr (1992) and found to give better results than simple integration at the midside nodes for this element.

The penalty factor is used to suppress the shear strains to model thin plates and using a mesh dependent value given by

$$\beta = 125t^2/32\Delta \tag{7.21}$$

improved convergence is obtained, smaller penalty values in coarse meshes reducing their stiffness and thus increasing the solution magnitudes.

The form of Equation 7.21 is arrived at by considering the need to balance the ratio of the bending and shearing stiffnesses. Globally the ratio of these depends on

$$Gt^2/1.2EL^2 \tag{7.22}$$

where $G = 0.5E/(1 + v)$ is the shear modulus, L is the plate span and t its thickness, whereas the discretization error depends upon

$$(h/L)^N \tag{7.23}$$

where h is the mesh spacing and N is the order of the truncation error of the element stiffness matrix.

Pessimistically assuming $N = 2$, as for constant strain elements, and taking the penalty factor to be proportional to the ratio of Equations 7.22 and 7.23 tends to give a cancellation of the shear and discretization errors:

$$\beta = (\text{const.})(Gt^2/1.2EL^2)(L/h)^2 = (\text{const.})t^2/h^2 = (\text{const.})t^2/\Delta \tag{7.24}$$

and the value of the factor used in Equation 7.21 was found by numerical trials (Mohr, 1992).

7.3. Numerical results

Table 7.1 shows the results given by the 18 freedom thick plate element for deflection and stress maxima for blanket loaded square plates. m_{max} and Q_{max} are sampled at the vertices and midsides and numerical integration at the midsides is used (Mohr, 1992). Note that in thick plate problems, of course, $\beta = 1$ is used.

Table 7.1. Results for blanket loaded thick plates ($\nu = 0.2$).

Nodes/quadrant	$w_{max}D/qL^4$	m_{max}/qL^2	Q_{max}/qL^2
Simply supported slab, $t/L = 0.2$			
9	0.00359	0.0347	0.167
25	0.00470	0.0451	0.252
49	0.00478	0.0451	0.281
81	0.00479	0.0449	0.295
Exact	0.00478	0.0442	0.338
Thin plate	0.00406	0.0442	0.338
Clamped slab, $t/L = 0.1$			
9	0.00104	0.0150	0.167
25	0.00136	0.0355	0.289
49	0.00144	0.0421	0.331
81	0.00146	0.0453	0.354
Argyris & Sharpf		0.0443	0.365
Griemann & Lynne	0.00148	0.0475	
Thin plate	0.00126	0.0513	0.338

The results are satisfactory and can be improved using the integration data of Equations 7.20.

The consistent loads for force distributed over the element area are given by

$$\{q_c\} = q \int\int\{f\}dxdy \qquad (7.25)$$

and the interpolation functions can be exactly integrated using the formula

$$\int\int L_1^a L_2^b L_3^c dxdy = [a!b!c!/(a+b+c+2)!](2\Delta) \qquad (7.26)$$

which for the quadratic functions of Equations 7.6 gives

$$\{q_c\} = (q\Delta/3)\{0,0,0,1,1,1\} \qquad (7.27)$$

so that one third of the total load on the element is applied at its midside nodes.

Table 7.2 shows the results obtained with the element in a plate with $L/t = 10$ and with various boundary conditions, loads, meshes and penalty factors. Here CL = clamped edges, SS = simply supported edges, UDL = uniformly distributed load and PL = central point load.

Here the integration of Equation 7.20 is used and both a constant penalty factor with the value given by Equation 7.21 for the finest mesh and gradually increasing penalty factors are used to show the effect of the latter upon convergence.

Table 7.2. Results for thin plates

Nodes	β	$w_{max}D/qL^4$		$w_{max}D/PL^2$	
		CL/UDL	SS/UDL	CL/PL	SS/PL
9	5	0.000493	0.003551	0.003560	0.009313
25	5	0.000996	0.003870	0.004501	0.010745
49	5	0.001200	0.004102	0.005388	0.011640
9	0.31	0.001534	0.004324	0.007172	0.013226
25	1.25	0.001280	0.004081	0.005887	0.011906
49	2,813	0.001262	0.004113	0.005721	0.11845
81	5	0.001277	0.004269	0.005727	0.012185
Exact		0.001265	0.004062	0.005605	0.11600

The results are satisfactory and the element is as accurate as other elements with similar numbers of freedoms, though it is evident that results for the SS case might be improved with the use of a slightly larger penalty factor.

7.4. Conclusions

The triangular element described provides a good example of the use of natural strains, extending these to deal with shear strains, resulting in an elegant formulation.

With the use of the simple LST interpolation for each degree of freedom the element is easy to derive.

The element is thus readily able to be extended to develop a curved shell element in Chapter 8.

7.5. References

Argyris JH, Sharpf, Finite element theory of plates and shells including transverse shear strain effects, *IUTAM symposium on high speed computation for elastic structures*, vol. 1, Univ. de Liége, 1971.

Griemann, LF, Lynne PP, Finite element analysis of plate bending with transverse shear deformation, *Nuclear Engineering & Design* 14 (1970) 223.

Mohr GA, A triangular finite element for thick slabs, *Computers & Structures* 9 (1978) 595.

Mohr GA, *Finite Elements for Solids, Fluids, and Optimization*, Oxford University Press, Oxford 1992.

Pryor CW, Barker RM, Frederick D, Finite element bending analysis of Reissner plates, *J. Engineering Mechanics. Division, ASCE* (American Society of Civil Engineers) 96 (1970) 974.

Pugh EDL, Hinton E, Zienkiewicz OC, A study of quadrilateral plate bending elements with reduced integration, *Int. J. Numerical Methods in Engineering* 12 (1978) 1059.

Rao, GV, Venkataramana J, Raju IS, A high precision triangular plate bending element for the analysis of thick plates, *Nuclear Engineering & Design 30* (1974) 408.

Chapter 8

CURVED SHELL ELEMENTS

8.1. Curved shell elements

Several finite elements have been developed especially for the analysis of cylindrical shells (Gallagher, 1966; Connor & Brebbia, 1967; Cantin & Clough, 1968; Ashwell & Sabir, 1972), but these have limited applications.

For the analysis of general shell shapes flat elements such as that derived in Chapter 6 have been much used, along with three dimensional isoparametric elements in which simple constraints are used to coalesce the nodes at the neutral surface (Ergatoudis et al., 1968). Irons' SemiLoof element is an extension of this approach in which transverse shear constraints are also included (Irons 1976).

Many doubly curved shell elements for the analysis of general shell shapes have been developed (Cowper et al., 1970; Dawe, 1975; Gallagher & Thomas, 1976; Mohr, 1980; Hughes, 1987), some of these requiring as many as 63 freedoms (Argyris & Scharpf, 1968).

The strains and curvatures in thin shells are given by equations of the form (Love, 1944)

$$\varepsilon_x = \partial u/\partial x + w/R_x, \quad \varepsilon_y = \partial v/\partial y + w/R_x$$

$$\varepsilon_{xy} = \partial u/\partial y + \partial v/\partial x + 2w/R_{xy}$$

$$\chi_x = -\partial^2 w/\partial x^2 + \partial(u/R_x)/\partial x$$

$$\chi_y = -\partial^2 w/\partial x^2 + \partial(v/R_y)/\partial y \quad (8.1)$$

$$\chi_{xy} = 2\partial^2 w/\partial x \partial y + \partial(u/R_x)/\partial y + \partial(v/R_y)/\partial x$$

where R_s and R_y are the radii of curvature and R_{xy} is the twist curvature of the surface.

The terms of the form w/R are the *radial strain* components and the u,v contributions to the curvatures take account of the effect of these displacements upon the slope of the tangent to the shell surface (Mohr, 1992), the latter terms having a relatively small effect (Mohr, 1992).

Numerical implementation of these equations is relatively straightforward, given suitable thin plate and plane stress elements to act as a basis for them. Analysis of the differential geometry of arbitrary shell shapes to determine the radii of curvature, however, is difficult. In addition we require such analysis to determine the curved area, side lengths etc. for finite elements of the shell surface. In the following section the *natural* approach is used to accomplish this.

8.2. A natural doubly curved shell element

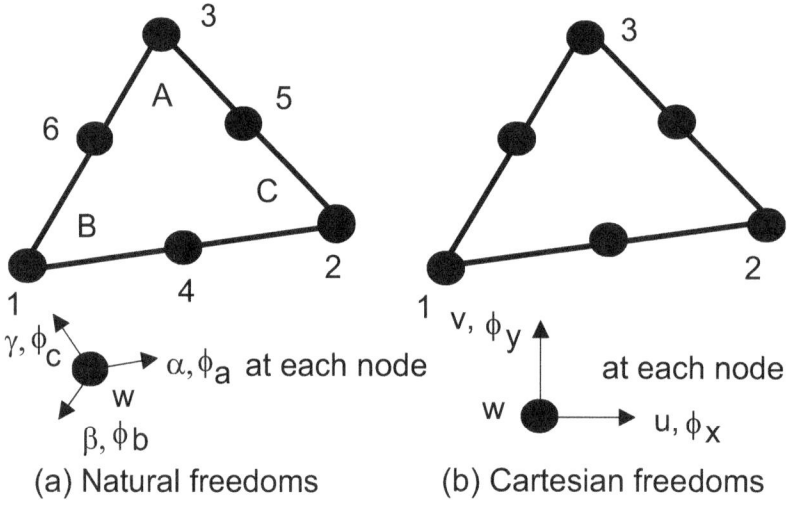

(a) Natural freedoms   (b) Cartesian freedoms

Figure 8.1. Doubly curved shell element.

Figure 8.1 shows the local Cartesian and natural freedoms used for a 30 freedom doubly curved shell element. Essentially the element combines the thick plate element of Section 7.2 with the classical linear strain triangle of Section 3.4.

Expressing Equations 8.1 in terms of natural coordinates parallel to the sides of a triangular element and including the average transverse shear strains (Naghdi, 1963)

$$\chi_a = (\partial a/\partial a)/R_a - \partial \phi_a/\partial s_a$$
$$\gamma_a = \partial w/\partial a - \phi_a + a/R_a \qquad\qquad (8.2)$$
$$\varepsilon_a = \partial a/\partial a + w/R_a \qquad\qquad a \rightarrow b, c$$

where a, a, R_a and ϕ_a are respectively curvilinear displacement, curvilinear coordinate, radius of curvature and flexural rotation referred to side a of the element (that is that joining nodes 1 and 2). These equations are repeated for sides b and c as $a \rightarrow b, c$ denotes.

The displacements a and slopes ϕ_a are given by

$$a = c_{ax}u + c_{ay}v, \quad \phi_a = c_{ax}\phi_x + c_{ay}\phi_y \quad a \rightarrow b, c \tag{8.3}$$

where the direction cosines for the first side, for example, are given by

$$c_{ax} = (x_2 - x_1)/L_a, \quad c_{ay} = (y_2 - y_1)/L_a \tag{8.4}$$

and here x,y are curvilinear coordinates whose plane projections are the global Cartesian axes X, Y, Z and L_a is the curved arc length of the element side.

Using the quadratic interpolation for the LST element the 9×30 strain interpolation matrix is given by

$$B = \begin{bmatrix} B_b \\ B_s \\ B_m \end{bmatrix} = \begin{bmatrix} u \text{ columns} & v \text{ columns} & w \text{ columns} & \phi_x \text{ columns} & \phi_y \text{ columns} \\ c_{ax}\{f_a\}^t/R_a & c_{ay}\{f_a\}^t/R_a & 0 & -c_{ax}\{f_a\}^t & -c_{ay}\{f_a\}^t \\ c_{ax}\{f\}^t/R_a & c_{ax}\{f\}^t/R_a & \{f_a\}^t & -c_{ax}\{f\}^t & -c_{ay}\{f\}^t \\ c_{ax}\{f_a\}^t & c_{ay}\{f_a\}^t & \{f\}^t/R_a & 0 & 0 \end{bmatrix} a \rightarrow b, c \tag{8.5}$$

where

$$\{f\} = \{2L_1^2 - 1, \ 2L_2^2 - 1, \ 2L_3^2 - 1, \ 4L_1 L_2, \ 4L_2 L_3, \ 4L_3 L_1\} \tag{8.6}$$

$$\{f_a\} = \{\partial\{f\}/\partial L_i - \partial\{f\}/\partial L_j\}/L_a \quad a \rightarrow b, c \quad i = 1, 2, 3 \quad j = 2., 3, 1 \tag{8.7}$$

Then using Equations 3.46 and 7.17 to transform the natural strains to Cartesian strains the transformed strain-displacement matrix is given as

$$B = \begin{bmatrix} B_b{}^* \\ B_s{}^* \\ B_m{}^* \end{bmatrix} = \begin{bmatrix} C_N^{-1} B_b \\ T_s B_s \\ C_N^{-1} B_b \end{bmatrix} \tag{8.8}$$

Including a penalty factor β the final element stiffness matrix is given by three point numerical integration

$$k = \Sigma \, [B_b^t{}^* \, D_b \, B_b{}^* + \beta B_s^t{}^* D_s B_s{}^* + (12/t^2)B_m^t{}^* D_b \, B_m{}^*] \, (\varpi_i \Delta) \tag{8.9}$$

using integration at three internal points given by the areal coordinates

$$(L_1, L_2, L_3) = (4/6, 1/6, 1/6), \ (1/6, 4/6, 1/6), \ (1/6, 1/6.4/6) \tag{8.10}$$

with weights $\varpi_i = 1/3$ at each. In Equation 8.9 D_b is the upper left 3×3 part of matrix D of Equation 7.3 and D_s is the lower right 2×2 part of D.

The penalty factor is used to suppress the shear strains to model thin plates and using a mesh dependent value determined for the thick plate element of Section 7.2

$$\beta = 125t^2/32\Delta \tag{8.11}$$

The values of this are calculated within the element subroutine because, generally, in shell problems element sizes will vary considerably.

Finally an alternative transformation to that of Equation 3.46 can be used to transform from Cartesian to natural strains. This defines *independent natural strain components* $\varepsilon_a{}^*, \varepsilon_b{}^*, \varepsilon_c{}^*$

$$\{\varepsilon_N\} = \begin{Bmatrix} \varepsilon_a \\ \varepsilon_b \\ \varepsilon_c \end{Bmatrix} = \begin{bmatrix} 1 & \cos^2 C & \cos^2 B \\ \cos^2 C & 1 & \cos^2 A \\ \cos^2 B & \cos^2 A & 1 \end{bmatrix} \begin{Bmatrix} \varepsilon_a{}^* \\ \varepsilon_b{}^* \\ \varepsilon_c{}^* \end{Bmatrix} = W\{\varepsilon_N{}^*\} \tag{8.12}$$

where A, B, C are the vertex angles respectively at nodes 3,1,2, as shown in Figure 8.1.

The total natural strains are related to the Cartesian strains by

$$\{\varepsilon_N\} = \begin{Bmatrix} \varepsilon_a \\ \varepsilon_b \\ \varepsilon_c \end{Bmatrix} = \begin{bmatrix} c_{ax}^2 & c_{ay}^2 & 2c_{ax}c_{ay} \\ c_{bx}^2 & c_{by}^2 & 2c_{bx}c_{by} \\ c_{cx}^2 & c_{cy}^2 & 2c_{cx}c_{cy} \end{bmatrix} \begin{Bmatrix} \varepsilon_x \\ \varepsilon_y \\ \varepsilon_{xy} \end{Bmatrix} = C_N\{\varepsilon_{xy}\} \tag{8.13}$$

Then we have

$$\{\varepsilon_{xy}\} = C_N^{-1}\{\varepsilon_N\} = C_N^{-1}W\{\varepsilon_N^*\} = G\{\varepsilon_N^*\} \tag{8.14}$$

so that C_N^{-1} in Equation 8.8 is replaced by $G^{-1} = W^{-1}C_N$.

This transformation is numerically more accurate because the inverted matrix W is positive definite, that is it is diagonally dominant and also cannot be singular.

8.3. Differential geometry calculations

The element geometry data are the Cartesian coordinates X,Y,Z of the six nodes, taking the midside nodes to be halfway between those of the vertices in plan. The arc lengths of each side are calculated using one dimensional quadratic interpolation on each side.

For the first side, for example, interpolating its nodal elevations

$$Z = (s^2 - s)Z_1/2 + (1 - s^2)Z_4 + (s^2 + s)Z_3/2 \qquad s = -1 \to 1 \tag{8.15}$$

where s is a dimensionless coordinate along the plan projection of this side of the element. Differentiating with respect to the dimensionless coordinate a along this side

$$dZ/da = (dZ/ds)(ds/da) = (2/P_a)(dZ/ds)$$
$$= [(2s - 1)Z_1 - 4sZ_4 + (2s + 1)Z_3]/P_a \tag{8.16}$$

where

$$P_a = [(X_2 - X_1)^2 + (Y_2 - Y_1)^2]^{1/2} \tag{8.17}$$

is the plan length of this side.

Then the arc length of the side is given by one dimensional numerical integration

$$L_a = P_a \sum_{i=1}^{2} \varpi_i \sec(\theta_i) \text{ where } \sec(\theta_i) = [1 + (dZ/da)_i]^{1/2} \tag{8.18}$$

where θ_i is the slope of side a at each numerical integration point and Gauss quadrature at $s_i = \pm\sqrt{3}$ with $\omega_i = \frac{1}{2}$ is used.

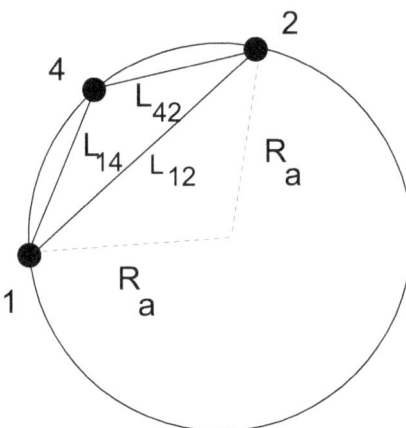

Figure 8.2. Calculation of curvature of element side.

To calculate the curvature of the element side a circle is fitted to its three nodes, as shown in Figure 8.2. Then from basic Euclidean geometry the curvature is given by

$$1/R_a = 4A_a/L_{14}L_{42}L_{12} \tag{8.19}$$

where A_a is the area of the triangle 142 in Figure 8.2, and this result is given a sign depending upon whether the midside node lies outside or inside the chord 12 between the end nodes.

Here the chord lengths L_{14}, L_{42}, L_{12} can easily be calculated using Pythagoras' theorem, for example

$$L_{12} = [(X_2 - X_1)^2 + (Y_2 - Y_1)^2 + (Z_2 - Z_1)^2]^{1/2} \tag{8.20}$$

but this considerably overestimates the curvatures when element sides are horizontal or nearly so.

This is because when the problem is that of a sphere, for example, and the element side under consideration is horizontal and near the crown of the shell, the curvature calculated using the lengths given by Equation 8.20 will be that for small circle, not a large circle for which the correct curvature applies.

An alternative approach which is used in the element routine given in the following section and which proves satisfactory for shallow shells is to assume that Figure 8.2 lies in a vertical plane, or approximately so.

Then the differences in elevation of the nodes in Figure 8.2 are calculated

$$Z_{14} = Z_1 - Z_4, \; Z_{42} = Z_4 - Z_2, \; Z_{12} = Z_1 - Z_2 \qquad (8.21)$$

and the chord lengths in Figure 8.2 are then calculated as

$$L_{14} = [P_a^2/4 + Z_{14}^2]^{1/2}, \; L_{42} = [P_a^2/4 + Z_{42}^2]^{1/2}, \; L_{12} = [P_a^2 + Z_{12}^2]^{1/2}$$
$$(8.22)$$

and the triangular area they enclose is given by the semi-perimeter formula

$$A_a = [p\,(p-L_{14})(p-L_{42})(p-L_{12})]^{1/2} \text{ where } p = (L_{14} + L_{42} + L_{12})/2$$
$$(8.23)$$

These arc length and curvature calculations were found more accurate than some alternative approaches and, notably, the use of natural strains and geometry avoids the considerable problem of calculating the twist curvature R_{xy} required by Equations 8.1.

8.4. Results for a cylindrical shell

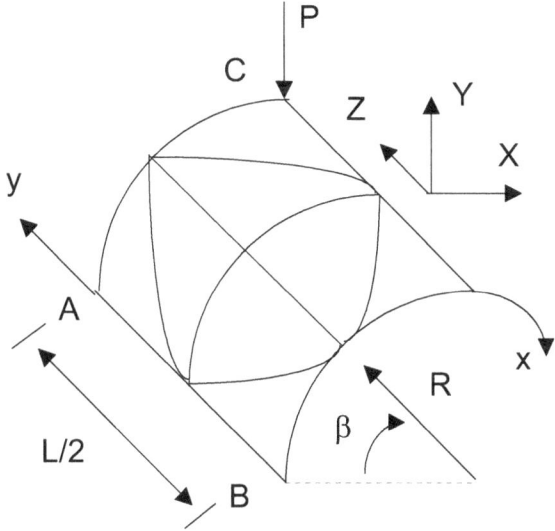

Figure 8.3. Pinch load cylinder with closed ends.
R/t = 100, L = 2R, υ = 0.3, P = 100, E = 10^6

Table 8.1 shows the results for the displacement under the load in the pinch loaded cylinder of Figure 8.3, an octant of which is analysed.

Original results using FORTRAN (Mohr, 1992) and recent results using BASIC (Mohr, 2003) are given.

The first two columns of results are obtained using specified local coordinates, that is one imagines a flat domain of size $\pi R/2$ in the curvilinear x direction and size L/2 in the y direction in Figure 8.3 and divides this into equal sized elements. Here an earlier version of this element which uses the Cartesian membrane strains of Equation 8.1 are used and the required curvatures are specified as the data $R_x = 0.1$, $R_y = 0$, $R_{xy} = 0$. Formulation for bending and shear, however, is the natural one of Section 8.2.

Table 8.1. *wEt/100P* under the load in Figure 8.3.

Elements/ nodes	Data Specified		Calculated	
	Fortran	BASIC	Fortran	BASIC
2/9	0.3462	0.3464	0.3381	0.3380
8/25	1.2269	1.2275	1.5688	1.5687
16/49	1.6337	1.6340	1.6844	1.6844
32/81	1.7157	1.7171	1.7009	1.7008
Extrapolated	1.8221	1.8239	1.7221	1.7219
Exact		1.7340		

The next third and fourth columns of results are obtained with only the global nodal coordinates given as data and using Equations 8.18 and 8.19 to calculate the arc lengths and curvatures of the element sides. Thus here all the generalized strains are calculated as natural values using Equation 8.5

Finally extrapolated values are obtained using h^N extrapolation (Mohr & Medland, 1983)

$$w_\infty = w_j + (w_j - w_i)/[(h_i/h_j)^N - 1] \qquad (8.24)$$

assuming $N = 2$ appropriate for the present element.

The results are satisfactory and the fully natural formulation of Section. 8.3 gives slightly improved results (results columns 3 and 4).

8.5. BASIC curved shell element code

The Visual BASIC (VB5/6) code for the fully natural doubly curved shell element formulation of Section 8.2, including the element geometry calculations of Section 8.3, follows.

Coding follows the details of Sections. 8.2 and 8.3 closely.

This second subroutine is called to calculate the matrices for element 'n' by the first main subroutine (MAIN) which first reads the problem data.

The third subroutine, the solution routine, is almost identical to that given in Section 6.4 and the fourth and final subroutine, the stress calculation subroutine is also almost identical to that given in Section 6.4.

The numerical integration loop for the element stiffness matrix is used three times at the integration points and a further three times with zero weights to form the stress matrix for the vertex nodes, the midside nodes or for the '4/6' integration points which tends to give best results for this element. In the latter instance the stresses are associated with the node closest to each integration point and nodal averaging of these is still used.

```
[Attribute VB_Name = "Module2" - VB heading - not BASIC code]
DefSng A-H, O-Z: DefInt I-N
Sub elmats(n)
50 Dim S(30, 30), C(24, 30), D(9, 9), B(9, 30), XY(6, 3), F(6), TEMP(9, 30)
60 Dim CT(3, 2), DG(2, 2), CG(2, 3), DF(9, 9), V(3, 6), PS(3), R(3)
80 L = imat(n): E = prop(L, 1) * 10 ^ 6: P = prop(L, 2)
Rem Collect element properties
90 DENS = prop(L, 3): TH = prop(L, 4): UDL = prop(L, 5)
95 For M = 1 To 6: K = nop(n, M): XY(M, 1) = cord(K, 1): XY(M, 2) = cord(K, 2)
XY(M, 3) = cord(K, 3):  Next: Rem matrix of element coords
100 X21 = XY(2, 1) - XY(1, 1): X32 = XY(3, 1) - XY(2, 1): X13 = XY(1, 1) - XY(3, 1)
110 Y21 = XY(2, 2) - XY(1, 2): Y32 = XY(3, 2) - XY(2, 2): Y13 = XY(1, 2) - XY(3, 2)
Z21 = XY(2, 3) - XY(1, 3): Z32 = XY(3, 3) - XY(2, 3): Z13 = XY(1, 3) - XY(3, 3)
PS(1)=Sqr(X21 ^ 2+Y21^ 2): PS(2) = Sqr(X32 ^ 2+Y32 ^ 2): PS(3) = Sqr(X13 ^ 2+Y13 ^ 2)
A = Abs(X21 * Y32 - X32 * Y21): Rem = 2x plan area
TX = (Y32 * XY(1, 3) + Y13 * XY(2, 3) + Y21 * XY(3, 3)) / A: Rem tangents of element
TY = -(X32 * XY(1, 3) + X13 * XY(2, 3) + X21 * XY(3, 3)) / A: Rem vertex node plane
TX = Sqr(1 + TX * TX): TY = Sqr(1 + TY * TY)
CT(1, 1) = TX * X21 / Sqr(PS(1) * PS(1) + Z21 * Z21): Rem local direction cosines
CT(1, 2) = TY * Y21 / Sqr(PS(1) * PS(1) + Z21 * Z21): Rem of sides
CT(2, 1) = TX * X32 / Sqr(PS(2) * PS(2) + Z32 * Z32)
CT(2, 2) = TY * Y32 / Sqr(PS(2) * PS(2) + Z32 * Z32)
CT(3, 1) = TX * X13 / Sqr(PS(3) * PS(3) + Z13 * Z13)
CT(3, 2) = TY * Y13 / Sqr(PS(3) * PS(3) + Z13 * Z13)
t = Sqr(1 / 3): I = 0: J = 3: K = 1
For L = 1 To 3: I = I + 1: J = J + 1: K = K + 1: If K = 4 Then K = 1
TX = (2 * t - 1) * XY(I, 3) - 4 * t * XY(J, 3) + (2 * t + 1) * XY(K, 3): TX = TX / PS(I)
```

TY = (-2 * t - 1) * XY(I, 3) + 4 * t * XY(J, 3) - (2 * t - 1) * XY(K, 3): TY = TY / PS(I)
AZ = XY(J, 3) - XY(I, 3): BZ = XY(K, 3) - XY(J, 3): CZ = AZ + BZ
AA = Sqr(PS(I) * PS(I) / 4 + AZ * AZ): BB = Sqr(PS(I) * PS(I) / 4 + BZ * BZ)
CC = Sqr(PS(I) * PS(I) + CZ * CZ): SS = (AA + BB + CC) / 2
ZM = (XY(I, 3) + XY(K, 3)) / 2: If XY(J, 3) = ZM Then GoTo 112
A = Sqr(SS * (SS - AA) * (SS - BB) * (SS - CC))
 R(I) = 4 * A / (AA * BB * CC): Rem curvature of element side
If XY(J, 3) < ZM Then R(I) = -R(I)
GoTo 113
112 R(I) = 0
113 PS(I) = 0.5 * PS(I) * (Sqr(1 + TX * TX) + Sqr(1 + TY * TY)): Next
Rem: PS() = arc lengths of element sides
S21 = PS(1): S32 = PS(2): S13 = PS(3): SS = (S21 + S32 + S13) / 2
A = Sqr(SS * (SS - S21) * (SS - S32) * (SS - S13)): A = 2 * A
penf = 125 * TH * TH / (16 * A): AA = A * A: Rem calculate penalty factor
210 CG(1, 1) = CT(2, 2) * S21 * S32 - CT(3, 2) * S13 * S21
220 CG(1, 2) = CT(3, 2) * S32 * S13 - CT(1, 2) * S21 * S32
230 CG(1, 3) = CT(1, 2) * S13 * S21 - CT(2, 2) * S32 * S13: Rem shear trans. matrix
240 CG(2, 1) = -CT(2, 1) * S21 * S32 + CT(3, 1) * S13 * S21
250 CG(2, 2) = -CT(3, 1) * S32 * S13 + CT(1, 1) * S21 * S32
260 CG(2, 3) = -CT(1, 1) * S13 * S21 + CT(2, 1) * S32 * S13
270 For I = 1 To 3: NI = I + 3: NF = nop(n, NI) * ndf - 2
280 q(NF) = q(NF) + A * UDL / 6: Next: Rem Loads for blanket loading
285 Rem ***** Form modulus matrix - for stress calcs
286 Rem ***** form DF which omits premultipications
290 D(1, 1) = E * TH ^ 3 / (12 * (1 - P * P)): D(2, 2) = D(1, 1)
D(1, 2) = P * D(1, 1): D(2, 1) = D(1, 2)
300 D(3, 3) = 0.5 * (1 - P) * D(1, 1): D(1, 3) = 0: D(2, 3) = 0: D(3, 1) = 0: D(3, 2) = 0
330 DG(1,1) =5 *E* TH / (12*(1+P)): DG(2, 2) = DG(1, 1): DG(1, 2) =0: DG(2, 1) =0
335 Rem ***** Form matrix C
340 B(1, 1) = CT(1, 1) ^ 2: B(1, 2) = CT(1, 2) ^ 2: B(1, 3) = 2 * CT(1, 1) * CT(1, 2)
350 B(2, 1) = CT(2, 1) ^ 2: B(2, 2) = CT(2, 2) ^ 2: B(2, 3) = 2 * CT(2, 1) * CT(2, 2)
360 B(3, 1) = CT(3, 1) ^ 2: B(3, 2) = CT(3, 2) ^ 2: B(3, 3) = 2 * CT(3, 1) * CT(3, 2)
370 AX = 1: BB = 1: CZ = 1: AZ = (S21 ^ 2 + S13 ^ 2 - S32 ^ 2) / (2 * S21 * S13)
380 AZ = AZ * AZ: CX = AZ: DD = 1 - AZ: AY = 1 - DD * (S13 / S32) ^ 2
390 BX = AY: BZ = 1 - DD * (S21 / S32) ^ 2: CY = BZ:Rem matrix W
Rem: calculate inverse of W:
400 DD =AX *(BB * CZ-BZ * CY) - AY *(BX * CZ-BZ * CX) + AZ *(BX * CY-BB * CX)
410 S(1, 1) = BB*CZ - BZ*CY: S(1, 2) = -(AY*CZ - AZ*CY): S(1, 3) = AY*BZ - AZ*BB
420 S(2, 1) = -(BX*CZ - BZ*CX): S(2, 2) = AX*CZ - AZ*CX: S(2, 3) = -(AX*BZ -AZ*BX)
430 S(3, 1) = BX*CY - BB*CX: S(3, 2) = -(AX*CY - AY*CX): S(3, 3) = AX*BB - AY*BX
440 For I = 1 To 3: For J = 1 To 3: C(I, J) = 0: For K = 1 To 3:
450 C(I, J) = C(I, J) + D(I, K) * B(J, K): Next: Next: Next:
460 For I = 1 To 3: For J = 1 To 3: DF(I, J) = 0: D(I, J) = 0: For K = 1 To 3
470 DF(I, J) = DF(I, J) + C(I, K) * S(K, J) / DD: D(I, J) = D(I, J) + B(I, K) * C(K, J)
480 Next: Next: Next:

```
490 For I = 1 To 3: For J = 1 To 3: C(I, J) = 0: For K = 1 To 3
500 C(I, J) = C(I, J) + D(I, K) * S(K, J) / DD: Next: Next: Next
510 For I = 1 To 3: For J = 1 To 3: D(I, J) = 0: For K = 1 To 3
520 D(I, J) = D(I, J)+S(I, K)*C(K, J)/DD: Next:Next:Next: Rem: transformed D matrix
For I = 1 To 3: For J = 1 To 3: DF(I + 6, J + 6) = DF(I, J) * 12 / (TH * TH)
D(I + 6, J + 6) = D(I, J) * 12 / (TH * TH): Next: Next: Rem factor D for plane stress
530 For I = 1 To 2: For J = 1 To 3: S(I, J) = 0: For K = 1 To 2
540 S(I, J) = S(I, J) + DG(I, K) * CG(K, J) / (9 * AA): Next: Next: Next
550 For I = 1 To 3: IA = I + 3: For J = 1 To 3: JA = J + 3: For K = 1 To 2
560 D(IA, JA) = D(IA, JA)+ CG(K, I)*S(K, J): Next:Next:Next: Rem shear mod. matrix
570 For I = 1 To 2: For J = 1 To 3: DF(I + 3, J + 3) = S(I, J) * A * 3: Next: Next:
580 For I = 1 To 30: For J = 1 To 30: S(I, J) = 0: Next: Next
590 For II = 1 To 6: Rem COMMENCE ELEMENT INTEGRATION LOOP ########
600 F1 = 4 * ci(II, 1): F2 = 4 * ci(II, 2): Rem Integration point coords
610 For I = 1 To 9: For J = 1 To 30: B(I, J) = 0: Next: Next: Rem Initialize B matrix
620 C1 = ci(II, 1): C2 = ci(II, 2): CC = C1 + C2
Rem: quadratic interpolation:
630 F(1) = 2*C1*C1 - C1: F(2) = 2*C2*C2 - C2: F(3) = 1 - 3*CC + 2*CC*CC
640 F(4) = 4*C1*C2: F(5) = 4*C2*(1 - CC): F(6) = 4*C1*(1 - CC): F3 = 4 - F1 - F2
650 For I = 1 To 3: For J = 1 To 6: V(I, J) = 0: Next: Next
660 V(1, 1) = F1 - 1: V(1, 4) = F2: V(1, 6) = F3: Rem d(f)/dL1
670 V(2, 2) = F2 - 1: V(2, 4) = F1: V(2, 5) = F3: Rem d(f)/dL2
680 V(3, 3) = F3 - 1: V(3, 5) = F2: V(3, 6) = F1: Rem d(f)/dL3
690 For J = 1 To 6: JA = 5 * J - 2
700 B(4, JA) = (V(2, J) - V(1, J)) / S21: Rem d(f)/da, d(f)/db, d(f)/dc for B matrix
710 B(5, JA) = (V(3, J) - V(2, J)) / S32
720 B(6, JA) = (V(1, J) - V(3, J)) / S13
730 Next
740 For I = 1 To 3: For J = 1 To 6: IA = I + 3: JA = 5 * J - 1
750 B(IA, JA) = B(IA, JA) - CT(I, 1) * F(J): Rem Entries in rows 4,5,6
760 B(IA, JA + 1) = B(IA, JA + 1) - CT(I, 2) * F(J)
JA = JA - 3: B(IA, JA) = B(IA, JA) + CT(I, 1) * F(J) * R(I)
B(IA, JA + 1) = B(IA, JA + 1) + CT(I, 2) * F(J) * R(I): Next: Next
770 For J = 1 To 6: JX = 5 * J - 1: JY = JX + 1
780 B(1, JX) = -CT(1, 1) * (V(2, J) - V(1, J)) / S21
790 B(1, JY) = -CT(1, 2) * (V(2, J) - V(1, J)) / S21
800 B(2, JX) = -CT(2, 1) * (V(3, J) - V(2, J)) / S32: Rem Entries in rows 1,2,3
810 B(2, JY) = -CT(2, 2) * (V(3, J) - V(2, J)) / S32: Rem for phi columns
820 B(3, JX) = -CT(3, 1) * (V(1, J) - V(3, J)) / S13
830 B(3, JY) = -CT(3, 2) * (V(1, J) - V(3, J)) / S13
JA = JX - 3: JB = JY - 3: B(7, JA) = -B(1, JX): B(7, JB) = -B(1, JY)
B(8, JA) = -B(2, JX): B(8, JB) = -B(2, JY): B(9, JA) = -B(3, JX): B(9, JB) = -B(3, JY)
B(1, JA) = B(1, JX) * R(1): B(1, JB) = B(1, JY) * R(1)
B(2, JA) = B(2, JX) * R(2): B(2, JB) = B(2, JY) * R(2)
B(3, JA) = B(3, JX) * R(3): B(3, JB) = B(3, JY) * R(3)
840 Next
```

For J = 1 To 6: JA = 5 * J - 2: B(7, JA) = F(J) * R(1)
B(8, JA) = F(J) * R(2): B(9, JA) = F(J) * R(3): Next: Rem radial strains
910 IA = 4 * (II - 1) + 1
920 For J = 1 To 30: C(IA, J) = 0: C(IA + 1, J) = 0: C(IA + 2, J) = 0: C(IA + 3, J) = 0
930 For K = 1 To 9
940 C(IA, J) = C(IA, J) + DF(1, K) * B(K, J)
950 C(IA + 1, J) = C(IA + 1, J) + DF(2, K) * B(K, J): Rem Stress matrix to calculate
960 C(IA + 2, J) = C(IA + 2, J) + DF(7, K) * B(K, J): Rem Mx,My,Nx,Ny at each point
970 C(IA + 3, J) = C(IA + 3, J) + DF(8, K) * B(K, J)
980 Next K: Next J
990 If wf(II) = 0 Then GoTo 1070
1000 For I = 1 To 9: For J = 1 To 30: TEMP(I, J) = 0: For K = 1 To 9
1010 TEMP(I, J) = TEMP(I, J) + D(I, K) * B(K, J): Next: Next: Next
1020 SMUL = wf(II) * A
1030 For I = 1 To 30: For J = 1 To 30: For K = 1 To 3
1040 S(I, J) = S(I, J) + B(K, I) * TEMP(K, J) * SMUL
1050 S(I, J) = S(I, J) + B(K + 3, I) * TEMP(K + 3, J) * SMUL * penf: Rem ESM
1060 S(I, J) = S(I, J) + B(K + 6, I) * TEMP(K + 6, J) * SMUL
1065 Next K: Next J: Next I
1070 Next II: Rem END OF ELEMENT INTEGRATION LOOP   ###########
1080 For I = 1 To 30: For J = 1 To 30
1090 Write #8, S(I, J): Next: Next: Rem FILE STIFFNESS MATRICES
1100 For I = 1 To 24: For J = 1 To 30
1110 Write #9, C(I, J): Next: Next: Rem FILE STRESS MATRICES
1115 Debug.Print n: Rem REPORT COMPLETION FOR ELEMENT
1130 End Sub

The data file (the RHS comments are not in the file) for the 9 node mesh for the problem of Figure 8.3 is:

```
9,2,8,1,35    # nodes, elements, b.c. nodes, ele. types, half band width
5,-4          penalty factor - over written in element routine
1,0.3,0,1,0   Young's modulus, Poisson's ratio, not used, thickness, UDL
10,0,0        curvatures x 1000 - calculated in element routine
0,0           global x-y coordinates, z coords calculated from Rx from above
0,50
0,100
50,0
50,50
50,100
100,0
100,50
100,100
```

1,7,3,4,5,2	element node numbers
3,7,9,5,8,6	
1,10110	boundary condition flags for b.c. nodes
2,10010	
3,11011	
4,10110	
6,01001	
7,10110	
8,10010	
9,11011	
9,0,0,-25,0,0	¼ of point central point load in Fig. 8.3
0,0,0,0,0,0	zero load data line to terminate data reading

As noted, the main subroutine which reads this data is similar to that given in Section 6.4.

8.6. Results for a spherical shell

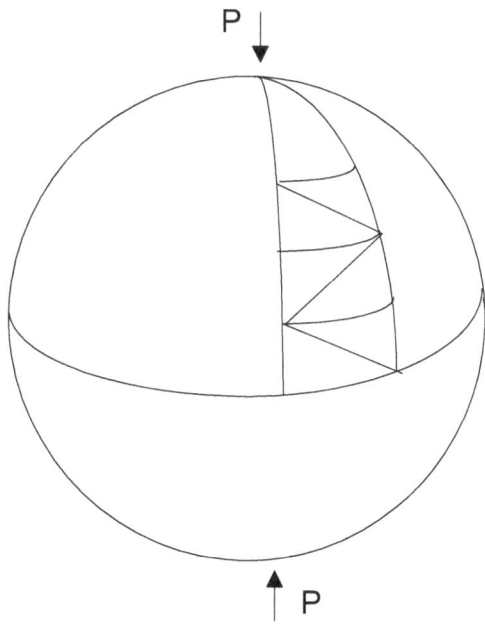

Figure 8.4. Pinch loaded spherical shell,
$E = 1$, $\upsilon = 0.3$, $t = 1$, $R = 50$, $P = 1$.

Fig. 8.4 shows a sector of a pinch loaded spherical shell modeled by a coarse mesh of seven doubly curved finite elements.

The nodal coordinate data are the spherical coordinates a, β such that

$$z = R\sin\beta, \quad R_p = R\cos\beta, \quad x = R_p \cos a, \quad y = R_p \sin a$$

$$(8.25)$$

In the element routine the resulting nodal coordinates are the only geometrical data. For this problem Equation 8.20 is used in the curvature calculation of Equation 8.19. As noted, for horizontal sides near the crown of the shell the small circle or latitudinal curvature is calculated. From this, of course, the large circle curvature can be calculated using Equations 8.25. Elsewhere, however, the curvatures of the element sides are calculated fairly accurately.

When a sector of the shell is analysed, as in Figure 8.4, present results suggest that the greater curvatures given by Equations 8.19 and 8.20 for horizontal sides near the crown better reflect the nature of the problem. Thus only three additional lines of code to calculate AA, BB, CC (the approximate side arc lengths) are added to the element routine following the original lines for this, and the calculation of the semi perimeter is SS moved down three lines to follow these new lines. Then the following line calculating ZM is removed (with a Rem) and the 10^6 multiplier for E is removed at the beginning of the element routine (line 80).

Note that the problem is solved in local coordinates, taking the origin at the crown and x^* being meridional and following a large circle bisecting the sector shown in Figure 8.4, y^* then being latitudinal. Thus the boundary conditions on the internal nodes along the meridional sides of the mesh sector have the boundary conditions 01001, that is $v^*, \partial v^*/\partial y^* = 0$.

Table 8.2. Results for spherical shell of Figure 8.4.

Nodes	Element sizes $\delta\beta$	-w at crown
24	10 + 20 + 2×30	81.6968
30	10 + 20 + 3 ×20	81.7032
36	3 ×10 + 3 ×20	82.0208
36	2 ×10 + 2 ×15 + 2 ×30	82.0238
42	5 ×10 + 2 ×20	82.0202
48	6 ×10 +2 ×15	82.0214
54	9 ×10	82.0221
Exact		84.80

Table 8.2 shows the results obtained for the displacement under the load using several meshes using a sector of angular width $\delta a = 20°$, agreeing well with the results of Table 6.5. Note, however, that different angular widths of the mesh sector yield different solutions.

The meshes used are denoted by the values of $\delta \beta$ used by the six node elements as they are used in pairs (except at the crown as shown in Figure 8.4) as they are deployed moving down from the crown to the equator. Note too that the use of smaller elements near the crown of the shell will stiffen it owing to the 'small circle' problem noted above.

The present results, therefore, involve approximations regarding curvature calculations for horizontal element sides near the crown, use of a limited number of elements near the crown, and use of a particular angular sector width for the meshes.

'Sector' meshes such as that used here have been used in other work on spherical shells, for example by Gallagher and Thomas (1976) who note the difficulties of modeling the rigid body motions and stress distributions with general shell elements in such problems.

8.7. Conclusions

The problem of numerically calculating curvatures accurately for general shell shapes, and doing so in a manner which will yield satisfactory results in the strain-displacement equations for doubly curved shells, remains a difficult one. Therefore, whilst the natural approach used here is perhaps the most promising, and certainly the most elegant approach to formulation of doubly curved shell elements, the facet shell element program given in Chapter 6 remains the simplest and most reliable approach and is at least as accurate.

8.8. References

Argyris JH, Scharpf, The SHEBA family of shell elements for the matrix displacement method, *Aeronautical Journal* 72 (1968) 873.

Ashwell DG, Sabir AB, A new cylindrical shell finite element based on simple independent strain functions, *Int. J. Numerical Methods in Engineering* 14 (1972) 171.

Cantin G, Clough RW, A curved cylindrical shell finite element, *J. American Institute of Aeronautics & Astronautics* 6 (1968) 1057.

Connor JJ, Brebbia CA, A stiffness matrix for a shallow rectangular shell element, *J. Engineering Mechanics Div., ASCE* 93 (1967) 43.

Cowper GR, Lindberg GM, Olson MD, A shallow shell finite element of triangular shape, *Int. J. Solids Structures* 6 (1970) 1133.

Dawe DJ, High order triangular finite element for shell analysis, *Int. J. Solids Structures* 11 (1975) 1097.

Ergatoudis, Irons BM, Zienkiewicz OC, Three dimensional analysis of arch dams and their foundations, *Symposium on Arch Dams, Instn Civil Engineers, London* 1968.

Gallagher RH, *The Development and Evaluation of Matrix Methods for Thin Shell Analysis*, PhD thesis, University of New York, Buffalo, 1966.

Gallagher RH, Thomas GR, A triangular element based on generalized potential energy concepts, in *Finite Elements for Thin Shells and Curved Members* (eds DG Ashell, RH Gallagher), Wiley, London 1976.

Hughes TJR, *The Finite Element Method: Linear Static and Dynamic Finite Element Analysis*, Prentice-Hall, Englewood Cliffs NJ, 1987.

Irons BM, SemiLoof shell element, in *Finite Elements for Thin Shells and Curved Members* (eds DG Ashwell, RH Gallagher), Wiley, London 1976.

Love AEH, *A Treatise on the Mathematical Theory of Elasticity*, 4th edn, Dover, NY NY, 1944.

Mohr GA, Numerically integrated triangular element for doubly curved thin shells, *Computers & Structures* 11 (1980) 565 – 571

Mohr GA, A doubly curved isoparametric shell element, *Computers & Structures* 14 (no. 1-2, 1981) 9 – 13.

Mohr GA, Application of penalty functions to a curved isoparametric thick shell element, *Computers & Structures* 15/6 (1982) 685-690.

Mohr, Doubly curved shell element using natural strain and geometry calculations, *Int. J. Arts & Sciences* 3 (2003) 15.

Naghdi PM, Foundations of elastic shell theory. In: *Progress in Mechanics IV* (eds Sheddon IM, Hill R), North-Holland, Amsterdam 1963.

Chapter 9

NATURAL ELEMENT FOR POTENTIAL FLOW

9.1. Potential flow

Potential flow refers to inviscid irrotational flow characterized by a potential function φ such that the velocities of flow are given by (Lamb, 1924)

$$\{u\} = -\nabla\phi \text{ or } u = -\partial\phi/\partial y, \quad v = -\partial\phi/\partial x \qquad (9.1)$$

Equations 9.1 identically satisfy the zero vorticity or irrotationality condition

$$\varpi_z = (\partial v/\partial x - \partial u/\partial y) = 0 \qquad (9.2)$$

Alternatively one defines an orthogonal stream function ψ such that

$$u = \partial\psi/\partial y, \quad v = -\partial\psi/\partial x \qquad (9.3)$$

Equations 9.3 identically satisfy the two dimensional continuity condition

$$\nabla\cdot\{u\} = \partial u/\partial x + \partial v/\partial y = 0 \qquad (9.4)$$

and lines of constant ψ are streamlines.

Both φ and ψ identically satisfy LaPlace's equation $\nabla^2() = 0$ which is seen by substituting Equations 9.1 into Equation 9.4

$$-\nabla\cdot\{u\} = \nabla\cdot\nabla\phi = \nabla^2\phi = \partial^2 u/\partial x + \partial^2 v/y^2 = 0 \qquad (9.5)$$

and substituting Equations 9.3 into Equation 9.2 gives $\nabla^2\psi = 0$ as the governing differential equation.

There are many examples of both types of problem, φ corresponding to temperature T in heat conduction and to gravitational head H in plane seepage, and ψ corresponding to Prandtl's stress function in plane torsion.

For potential flow problems the general problem is governed by Equation 9.5 subject to Neumann boundary conditions $\partial\phi/\partial n = 0$ on impermeable boundaries and $\partial\phi/\partial n = V_n$ elsewhere, as shown in Figure 9.1.

To form element equations the Galerkin method is used, that is weighting the governing differential equation at the sampling points with the same interpolation as used for its variables. This is equivalent to the virtual work method in the mechanics of solids because, like energy methods leading to element matrices given by the congruent transformation $k = B'DB$, it gives element matrices with the same least squares character (Mohr, 1992).

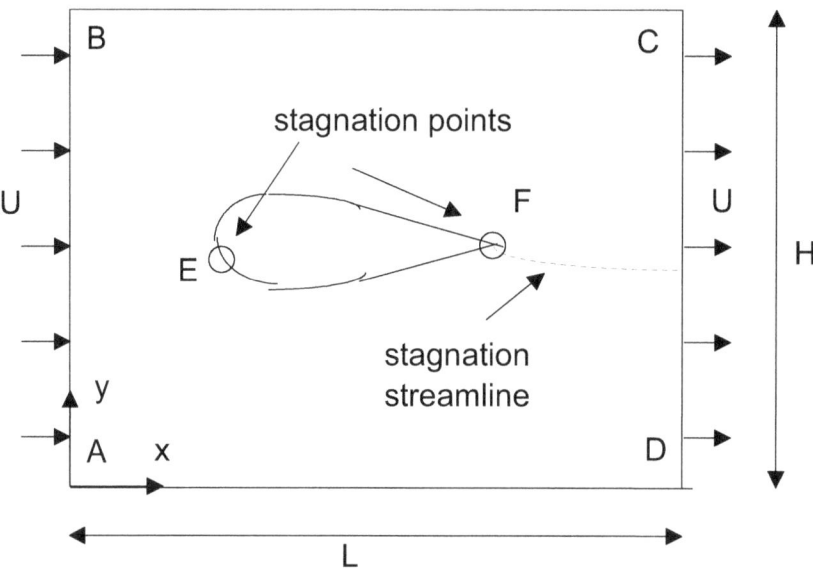

Figure 9.1. Potential flow past an aerofoil (Argyris & Dunne, 1976).

Boundary conditions for the surfaces are:

AB: $\phi = UL$, $\partial\phi/\partial x = U$, $\psi = Uy$, $\partial\psi/\partial x = 0$, BC: $\partial\phi/\partial y = 0$, $\psi = UH$

CD: $\phi = 0$, $\partial\phi/\partial x = 0$, $\psi = Uy$, $\partial\psi/\partial x = 0$, DA: $\partial\phi/\partial y = 0$, $\psi = 0$

Introducing interpolation functions $\{f\}$ for ϕ and then applying Galerkin weighting to Equation 9.5, integration by parts is then used to reduce the order of the governing equations and create boundary loading terms (Mohr, 1992).

For the first term, assuming an element of unit thickness, integration by parts gives

$$\iint \{f\}\{\partial^2 f/\partial x^2\}^t dxdy = \int \{f\}\{\partial f/\partial x\}^t dy - \iint \{\partial f/\partial x\}\{\partial f/\partial x\}^t dxdy$$

so that we obtain

$$[\ \iint(\{f_x\}\{f_x\}^t + \{f_y\}\{f_y\}^t)dxdy\ ]\{\phi\} = k\{\phi\} = \{q\} \tag{9.6}$$

where

$$\{q\} = \int\{f\}[c_x(\partial\phi/\partial x) + c_y(\partial\phi/\partial y)]dS = -\int\{f\}(uc_x + vc_y)dS \tag{9.7}$$

where c_x, c_y are the direction cosines of the outward directed normal to boundary and

$$\{f_x\} = \partial\{f\}/\partial x, \quad \{f_y\} = \partial\{f\}/\partial y \tag{9.8}$$

and Equations 9.1 are used to yield the final results for the velocity flux forcing terms at the boundaries. Without any Dirichlet boundary conditions for φ, however, the system equations are singular and thus $\varphi = 0$ must be specified as a datum value for at least one node.

For stream function formulations ψ simply replaces φ in Equations 9.6 and 9.7 but the final RHS in Equation 9.7 becomes

$$\{q\} = \int\{f\}(u\,c_y - v\,c_x)dS \tag{9.9}$$

and this vanishes when the resultant velocity flux F is perpendicular to the inlet boundary, that is when $u = Fc_x$, $v = Fc_y$ as in Figure 9.1, so that such problems require only the simple boundary conditions for ψ shown.

Figure 9.1 shows typical potential and stream function boundary conditions, setting datum values on one boundary and using Equations 9.1 and 9.3 to obtain the other boundary values:

$$\psi = \int(\partial\psi/\partial y)dy = Uy \tag{9.10a}$$

is used to find values at the inlet, outlet and upper boundaries, whilst

$$\phi = \int(\partial\phi/\partial x)dx = -Ux \tag{9.10b}$$

is used to obtain values at the inlet boundary.

Note, however, that Equation 9.11 only applies if the inlet boundary is relatively remote from the aerofoil, otherwise the safest course is to specify only the datum values at outlet and use the fluxes given by Equation 9.7 to force the flow.

When the distribution of φ or ψ has been obtained the pressure distribution can be found using Bernoulli's theorem (Mohr, 1992) and, in the example of Figure 9.1, the pressure is then integrated around the aerofoil to determine the aerodynamic lift.

9.2. Cubic element for potential flow

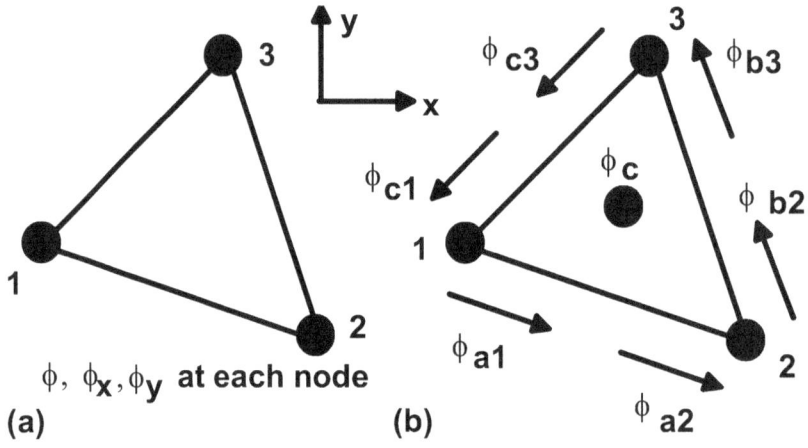

Figure 9.2. (a) 9 df global element, (b) 10 df local element.

Figure 9.2(a) shows the global freedoms for a nine freedom element for potential flow, namely the potential ϕ and its Cartesian derivatives $\phi_x = \partial\phi/\partial x$, $\phi_y = \partial\phi/\partial y$ at each node (Mohr and Power, 2003).

Defining *natural derivatives*

$$\phi_a = \partial\phi/\partial a, \ \phi_b = \partial\phi/\partial b, \ \phi_c = \partial\phi/\partial c \tag{9.11}$$

parallel to the element sides the local freedoms are shown in Figure 9.2(b), these natural derivatives corresponding to *natural velocities* of flow parallel to the element sides.

In order to express the interpolation consistently in terms of dimensionless interpolation functions the natural derivatives are expressed as dimensionless derivatives,

$$\phi_a{}^* = L_a(\partial\phi/\partial a) = x_{21}\phi_x + y_{21}\phi_y \text{ and similarly for } \phi_b{}^*, \phi_c{}^*,$$

(9.12)

where $x_{21} = x_2 - x_1$, $y_{21} = y_2 - y_1$ etc. and a is a coordinate parallel to side 12. Note that $c_{ax} = x_{21}/L_a$, $c_{ay} = y_{21}/L_a$ are the direction cosines of side 1-2, L_a being its length.

Including a centroidal freedom ϕ_c the complete cubic interpolation in areal coordinates is that of Equations 3.29, that is

$$\phi = \{f\}^t \{\phi_1, \phi_2, \phi_3, \phi_{a1}{}^*, \phi_{c1}{}^*, \phi_{b2}{}^*, \phi_{a2}{}^*, \phi_{c3}{}^*, \phi_{b3}{}^*, \phi_c\} = \{f\}^t\{d\},$$

(9.13)

where

$$f_1 = L_1 + L_1^2 L_2 + L_1^2 L_3 - L_2^2 L_1 - L_3^2 L_1 - 9L_1 L_2 L_3 \quad (\text{for } \phi_1)$$

$$f_2 = L_2 + L_2^2 L_3 + L_2^2 L_1 - L_3^2 L_2 - L_1^2 L_2 - 9L_1 L_2 L_3 \quad (\text{for } \phi_2)$$

$$f_3 = L_3 + L_3^2 L_1 + L_3^2 L_2 - L_1^2 L_3 - L_2^2 L_3 - 9L_1 L_2 L_3 \quad (\text{for } \phi_3)$$

$$f_4 = L_1^2 L_2 - L_1 L_2 L_3, \quad f_5 = -L_1^2 L_3 + L_1 L_2 L_3 \quad (\text{for } \phi_{a1}{}^*, \phi_{c1}{}^*)$$

$$f_6 = L_2^2 L_3 - L_1 L_2 L_3, \quad f_7 = -L_2^2 L_1 + L_1 L_2 L_3 \quad (\text{for } \phi_{b2}{}^*, \phi_{a2}{}^*)$$

$$f_8 = L_3^2 L_1 - L_1 L_2 L_3, \quad f_9 = -L_3^2 L_2 + L_1 L_2 L_3 \quad (\text{for } \phi_{c3}{}^*, \phi_{b3}{}^*)$$

$$f_{10} = 27 L_1 L_2 L_3 \quad (\text{for } \phi_c)$$

and the areal coordinates are given by cyclic permutation of

$$L_1 = 1/3 - (y_{32}x - x_{32}y)/2\Delta$$

where $2\Delta = |x_{21}y_{32} - x_{32}y_{21}|$ gives the element area Δ.

To eliminate the inconvenient centroidal freedom Equation 4.14 is used, that is

$$\phi_c = (\phi_1 + \phi_2 + \phi_3)/3 + (\phi_{a1}{}^* - \phi_{c1}{}^* + \phi_{b2}{}^* - \phi_{a2}{}^* + \phi_{c3}{}^* - \phi_{b3}{}^*)/18$$

$$(9.14)$$

Substituting Equation 9.14 into the interpolation of Equations 9.13, that is into

$$\phi = {}_{i=1}\Sigma^9\ f_i d_i + 27\ L_1 L_2 L_3 \phi_c$$

the modified interpolation functions are immediately obtained as

$$f_i{}^* = f_i + 9L_1 L_2 L_3 \ (i = 1,2,3)$$

$$f_4{}^* = L_1^2 L_2 + L_1 L_2 L_3/2, \quad f_5{}^* = -L_1^2 L_3 - L_1 L_2 L_3/2$$

$$f_6{}^* = L_2^2 L_3 + L_1 L_2 L_3/2, \quad f_7{}^* = -L_2^2 L_1 - L_1 L_2 L_3/2$$

$$f_8{}^* = L_3^2 L_1 + L_1 L_2 L_3/2, \quad f_9{}^* = -L_3^2 L_2 - L_1 L_2 L_3/2$$

$$(9.15)$$

and these are the interpolation functions of the classical BCIZ element described in Section 4.1, the simple explicit interpolation of which is still attractive.

Using Equations 9.15, an interpolation matrix for the first derivatives is easily obtained as

$$\{ \partial()/\partial L_1,\ \partial()/\partial L_2,\ \partial()/\partial L_3 \} = B\{ d^* \} \tag{9.16}$$

where

$$B^t = \begin{bmatrix} 1 + L_1(2 - L_1) - S & L_1^2 - 2L_2 L_1 & L_1^2 - 2L_3 L_1 \\ L_2^2 - 2L_1 L_2 & 1 + L_2(2 - L_2) - S & L_2^2 - 2L_3 L_2 \\ L_3^2 - 2L_1 L_3 & L_3^2 - 2L_2 L_3 & 1 + L_3(2 - L_3) - S \\ 2L_1 L_2 + a & L_1^2 + b & c \\ -2L_1 L_3 - a & -b & -L_1^2 - c \\ a & 2L_2 L_3 + b & L_2^2 + c \\ -L_2^2 - a & -2L_2 L_1 - b & -c \\ L_3^2 + a & b & 2L_3 L_1 + c \\ -a & -L_3^2 - b & -2L_3 L_2 - c \end{bmatrix} \tag{9.17}$$

and $S = L_1^2 + L_2^2 + L_3^2$, $a = L_2 L_3/2$, $b = L_3 L_1/2$, $c = L_1 L_2$.

This relates to the local freedoms. Generalizing Equations 9.12, the transformation from the global freedoms to these is simply

$$\{ d^* \} = \{ \phi_1, \phi_2, \phi_3, \phi_{a1}^*, \phi_{c1}^*, \phi_{b2}^*, \phi_{a2}^*, \phi_{c3}^*, \phi_{b3}^* \}$$

$$= T\{ d \} = T\{ \phi_1, \phi_{x1}, \phi_{y1}, \phi_2, \phi_{x2}, \phi_{y2}, \phi_3, \phi_{x3}, \phi_{y3} \},$$

(9.18)

where

$$T = \begin{bmatrix}
1 & 0 & 0 & 0 & 0 & 0 & 0 & 0 & 0 \\
0 & 0 & 0 & 1 & 0 & 0 & 0 & 0 & 0 \\
0 & 0 & 0 & 0 & 0 & 0 & 1 & 0 & 0 \\
0 & x_{21} & y_{21} & 0 & 0 & 0 & 0 & 0 & 0 \\
0 & x_{13} & y_{13} & 0 & 0 & 0 & 0 & 0 & 0 \\
0 & 0 & 0 & 0 & x_{32} & y_{32} & 0 & 0 & 0 \\
0 & 0 & 0 & 0 & x_{21} & y_{21} & 0 & 0 & 0 \\
0 & 0 & 0 & 0 & 0 & 0 & 0 & x_{13} & y_{13} \\
0 & 0 & 0 & 0 & 0 & 0 & 0 & x_{32} & y_{32}
\end{bmatrix}.$$

(9.19)

The derivatives in Equations 9.16 are transformed to Cartesian derivatives

$$\begin{Bmatrix} \partial(\ )/\partial x \\ \partial(\ )/\partial y \end{Bmatrix} = (1/2\Delta) \begin{bmatrix} -y_{32} & -y_{13} & -y_{21} \\ x_{32} & x_{13} & x_{21} \end{bmatrix} \begin{Bmatrix} \partial(\ )/\partial L_1 \\ \partial(\ )/\partial L_2 \\ \partial(\ )/\partial L_3 \end{Bmatrix}$$

(9.20)

using the areal coordinate definitions given earlier, and the 2×3 matrix in Equation 9.20 we denote by Z.

Then the final element matrix for Laplacian problems $(\nabla^2 \phi = 0)$ is given by

$$k = \Sigma\, FF^t (\omega_i \Delta) \text{ where } F = ZBT$$

(9.21)

using the numerical integration of quartic accuracy with integration point coordinates and weights shown in the following table (Mohr, 1992):

Point	L_1	L_2	L_3	ω_i
1	1-2a	a	a	c
2	a	1-2a	a	c
3	a	a	1-2a	c
4	1-2b	b	b	1/3 - c
5	b	1-2b	b	1/3 - c
6	b	b	1-2b	1/3 - c
a = 0.09157 62135, b = 0.44594 84909, c = 0.10995 17436				

Finally, for the case of Poisson's equation $\nabla^2 \phi = C$, the 'local' consistent loads are given by explicit integration of Equations 9.15 using Equation 4.42, yielding

$$\{ q_c^* \} = (C\Delta)T^t \{ 8, 8, 8, 1, -1, 1, -1, 1, -1 \}/24,$$

$$(9.22)$$

where in the case of plane torsion $C = 2G\theta$. Using the simple matrix T of Equation 9.19 we obtain

$$q_{c1} = q_{c2} = q_{c3} = q\Delta/3$$

$$q_{c2} = q\Delta(x_{21} - x_{13})/24, \quad q_{c3} = q\Delta(y_{21} - y_{13})/24$$

$$q_{c5} = q\Delta(x_{32} - x_{21})/24, \quad q_{c6} = q\Delta(y_{32} - y_{21})/24$$

$$q_{c8} = q\Delta(x_{13} - x_{32})/24, \quad q_{c9} = q\Delta(y_{13} - y_{32})/24 \qquad (9.23)$$

These explicit formulae are particularly useful and can be used for other 9 df Hermitian triangular elements where simple formulae are not available.

The coding of the element is particularly simple and transformation by the matrix T in Equation 9.21 can, observing the simple form of Equation 9.19, be coded explicitly rather than as a matrix multiplication.

9.3. Formulation with Lagrangian basis

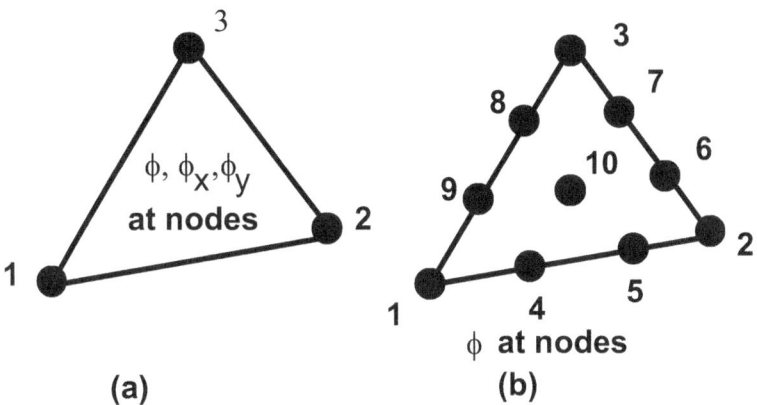

Figure 9.3. Cubic element freedoms: (a) global, (b) local.

Figure 9.3(b) shows the local freedoms for an alternative cubic element for potential flow. To transform to these we apply one dimensional cubic Hermitian interpolation on each side of the element to determine the ϕ values at the third points. On side 12, for example, the interpolation is

$$\phi = f_1\phi_1 + f_2(L_a\phi_{a1}) + f_3\phi_2 + f_4(L_a\phi_{a2}) \tag{9.24}$$

where

$$f_1 = 1 - 3s^2 + 2s^3, \quad f_2 = s - 2s^2 + s^3, \quad f_3 = 3s^2 - 2s^3. \quad f_4 = s^3 - s^2 \quad s = 0 \to 1 \tag{9.25}$$

and ϕ_{a1}, ϕ_{a2} are the (negative) natural velocities parallel to this side and these are obtained using Equations 3.20 and 3.21, yielding

$$\phi = f_1\phi_1 + f_2(x_{21}\phi_{a1} + y_{21}\phi_{a1}) + f_3\phi_3 + f_4(x_{21}\phi_{a2} + y_{21}\phi_{a2}) \tag{9.26}$$

At nodes 4 and 5 we have $s = 1/3$ and $s = 2/3$ and substituting these values into Equations 9.25 and using the results in Equation 9.26

$$27\phi_4 = 20\phi_1 + 4x_{21}\phi_{x1} + 4y_{21}\phi_{y1} + 7\phi_2 - 2x_{21}\phi_{x2} - 2y_{21}\phi_{y2} \tag{9.27a}$$

$$27\phi_5 = 7\phi_1 + 2x_{21}\phi_{x1} + 2y_{21}\phi_{y1} + 20\phi_2 - 4x_{21}\phi_{x2} - 4y_{21}\phi_{y2} \tag{9.27b}$$

117

Repeating this exercise on the other two sides the required basis transformation is

$$\{d_N\} = \{\phi_1.\phi_2, \phi_3 \ldots \phi_{10}\} = T\{\phi_1, \phi_{x1}, \phi_{y1} \ldots \phi_3, \phi_{x3}, \phi_{y3}\} = T\{d\}$$

(9.28)

where

$$T = (1/27)\begin{bmatrix} 27 & 0 & 0 & 0 & 0 & 0 & 0 & 0 & 0 \\ 0 & 0 & 0 & 27 & 0 & 0 & 0 & 0 & 0 \\ 0 & 0 & 0 & 0 & 0 & 0 & 27 & 0 & 0 \\ 20 & 4x_{21} & 4y_{21} & 7 & -2x_{21} & -2y_{21} & 0 & 0 & 0 \\ 7 & 2x_{21} & 2y_{21} & 20 & -4x_{21} & -4y_{21} & 0 & 0 & 0 \\ 0 & 0 & 0 & 20 & 4x_{32} & 4y_{32} & 7 & -2x_{32} & -2y_{32} \\ 0 & 0 & 0 & 7 & 2x_{32} & 2y_{32} & 20 & -4x_{32} & -4y_{32} \\ 7 & -2x_{13} & -2y_{13} & 0 & 0 & 0 & 20 & 4x_{13} & 4y_{13} \\ 20 & -4x_{13} & -4y_{13} & 0 & 0 & 0 & 7 & 2x_{13} & 2y_{13} \\ & & & \Sigma(\text{ rows } 4\text{-}9)/4 - \Sigma(\text{rows } 1-3)/6 & & & & & \end{bmatrix}$$

(9.29)

The bottom row of this matrix uses the approximation

$$w_{10} = (w_4 + w_5 + w_6 + w_7 + w_8 + w_9)/4 - (w_1 + w_2 + w_3)/6 \qquad (9.30)$$

and this was derived in Section 4.2.

Applying Equation 9.29 to the global freedoms of Figure 9.2(a) gives the local freedoms of Figure 9.2(b) and to these the standard cubic Lagrangian areal coordinate interpolation can be applied. These are Equations 3.17, but note that the node numbering of Figure 9.2(b) differs from that of Figure 3.3 and the interpolation functions are renumbered here accordingly:

$$f_1 = 4.5L_1^3 - 4.5L_1^2 + L_1, \quad f_2 = 4.5L_2^3 - 4.5L_2^2 + L_2, \quad f_3 = 4.5L_3^3 - 4.5L_3^2 + L_3$$
$$f_4 = 13.5L_1^2L_2 - 4.5L_1L_2, \quad f_5 = 13.5L_2^2L_1 - 4.5L_1L_2$$
$$f_6 = 13.5L_2^2L_3 - 4.5L_2L_3, \quad f_7 = 13.5L_3^2L_2 - 4.5L_2L_3$$
$$f_8 = 13.5L_3^2L_1 - 4.5L_3L_1, \quad f_9 = 13.5L_1^2L_3 - 4.5L_3L_1$$
$$f_{10} = 27L_1L_2L_3$$

(9.31)

The interpolation matrix B for the first derivatives is easily calculated as

$$\{\partial\phi/\partial L_1, \partial\phi/\partial L_2, \partial\phi/\partial L_3\} = B\{\phi_i\} \quad \bar{i} = 1 \to 10 \qquad (9.32)$$

where

$$B^t = \begin{bmatrix}
13.5L_1^2 - 9L_1 + 1 & 0 & 0 \\
0 & 13.5L_2^2 - 9L_2 + 1 & 0 \\
0 & 0 & 13.5L_3^2 - 9L_3 + 1 \\
27L_1L_2 - 4.5L_2 & 13.5L_1^2 - 4.5L_1 & 0 \\
13.5L_2^2 - 4.5L_2 & 27L_1L_2 - 4.5L_1 & 0 \\
0 & 27L_2L_3 - 4.5L_3 & 13.5L_2^2 - 4.5L_2 \\
0 & 13.5L_3^2 - 4.5L_3 & 27L_2L_3 - 4.5L_2 \\
13.5L_3^2 - 4.5L_3 & 0 & 27L_3L_1 - 4.5L_1 \\
27L_3L_1 - 4.5L_3 & 0 & 13.5L_1^2 - 4.5L_1 \\
27L_2L_3 & 27L_1L_3 & 27L_1L_2
\end{bmatrix}$$

$$(9.33)$$

The final element matrix is given by Equations 9.20 - 9.21 with B and T given by Equations 9.33 and 9.29, using the six point integration data given in Section 9.2.

The element gives identical results to the formulation of Section 9.2 because, as demonstrated in Section 4.2 their interpolations (including their transformation matrices T) are equivalent. Because it uses simpler Lagrangian interpolations and its T matrix allows the option of retaining the centroidal freedom, however, the present element is preferable.

9.4. A simple test problem

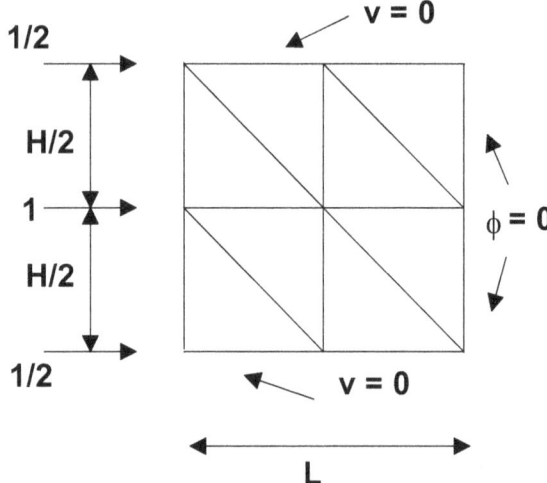

Figure 9.4. Rectilinear flow problem.

The element is first tested with the simple rectilinear potential flow problem of Figure 9.4.

To force the flow loads $q_\phi = 1/2, 1, 1/2$ are specified at inlet and $\phi = 0$ is set as a datum at outlet. It is also necessary to set $v = -\partial\phi/\partial y = 0$ at the top and bottom of the domain.

This yields the expected results

$$u = -\partial\phi/\partial x = 1, \quad \phi_{in} = QL/H, \tag{9.34}$$

where $Q = \Sigma\, q_\phi$ at the inlet.

9.5. Convergence of the element

The plane torsion problem is useful to test the convergence of the element. Here ϕ is a stress function such that

$$\tau_{zx} = \partial\phi/\partial y = -v, \quad \tau_{zy} = \partial\phi/\partial x = u \tag{9.35}$$

and the governing PDE is

$$\nabla^2\phi + 2G\theta = 0. \tag{9.36}$$

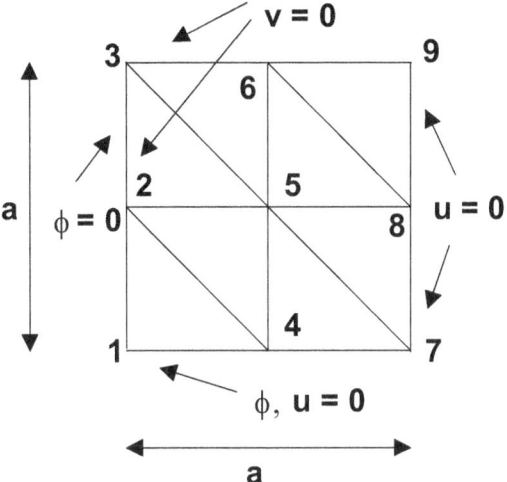

Figure 9.5. Plane torsion problem.

Figure 9.5 shows a quadrant of a square shaft section in plane torsion modeled using eight elements. The boundary conditions for the edges of the domain are shown, those at the corners combining those of the intersecting edges.

Corresponding to the constant term of Equation 9.36 'lumped' loads of

$$q_{\phi 5} = 2a^2/4, \quad q_{\phi 6} = q_{\phi 8} = (1/2)(2a^2/4), \quad q_{\phi 9} = (1/4)(2a^2/4)$$

$$(9.37)$$

are applied, in which $G\theta = 1$ is assumed.

The results for ϕ at the centre $(\phi^* = \phi_{max}/a^2)$ and τ at the middle of the sides $(\tau^* = \tau_{max}/2a^2)$ are compared to those of the 6 node quadratic Lagrangian element in Table 9.1.

Here meshes with 9 nodes and df (mesh 1) and 25 nodes and df (mesh 2) are used for the 6 df element and meshes with 4 nodes and 12 df (mesh 1) and 9 nodes and 27 df (mesh 2) are used for the new 9 df element.

Table 9.1. Plane torsion solutions (LL = lumped loads, CL = consistent loads).

Solution	Mesh 1	Mesh 2	Extrapolated	Basis
6 df, φ^*	0.6000	0.5900	0.5893	h^4
6 df, τ^*	0.6500	0.6568	0.6590	h^2
9 df, LL, φ^*	0.5357	0.5762	0.5897	h^2
9 df, LL, τ^*	0.2143	0.3016	n/a	n/a
9 df, CL, φ^*	0.5714	0.5888	0.5894	h^5
9 df, CL, τ^*	0.6786	0.6705	0.6702	h^5
Exact, φ^*			0.5894	
τ^*			0.6753	

The results are extrapolated using h^N extrapolation [Equation 8.24] and compared with those of the series solution. For ϕ^* we require $N = 2(p - m + 1) = 4$ and for τ^* we require $N = 2$ for the quadratic element [Mohr & Medland, 1983].

For the cubic element, $N = 2$ appears appropriate for ϕ^* owing to the crude lumped load approximation, but the results for τ^* are poor and cannot be extrapolated. The remedy is to use the consistent loads of Equation 9.23, when h^5 extrapolation is found appropriate for both ϕ^* and τ^*, because the latter is now a nodal freedom.

This is intermediate between $N = 4$ and $N = 6$ for quadratic and cubic elements, not uncommon for elements involving approximations such as Equation 9.14, and the results are clearly satisfactory.

Note that greater accuracy of stress or velocity solutions (with $N = 5$) is a principal advantage of the cubic element, one that is demonstrated in the following section.

9.6. A classical test problem

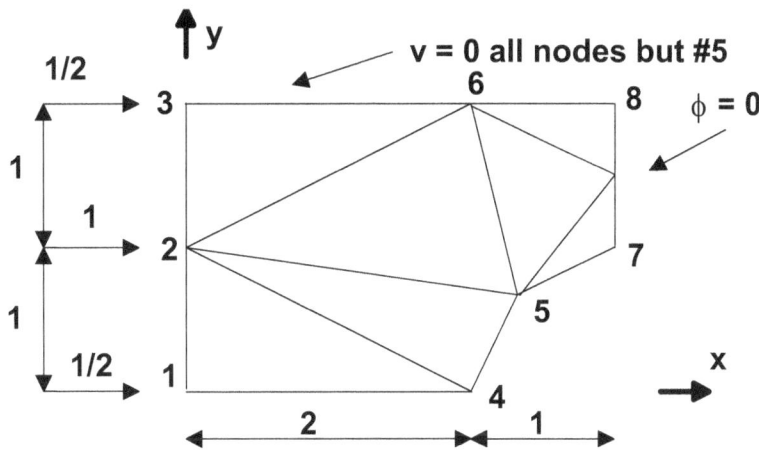

Figure 9.6. Flow around a cylinder.

Figure 9.6 shows the element applied to the classical problem of flow around a cylinder (Martin, 1968), analysing one quadrant. The loads q_ϕ shown are specified at the inlet and $\phi = 0$ is set at the outlet, with $v = 0$ around the boundary (except at node 5). The area 'cut out' by the cylinder is $A = \pi r^2/4 \simeq 0.8$, so that the coordinates of node 5 are chosen as $x = 2.2$ and $y = 0.8$. This gives

$$A = 1 - 2(1/2)(1)(0.2) = 0.8 \tag{9.38}$$

yielding a reasonable approximation of the boundary shape.

The results for the velocity over the crest of the cylinder are compared in Table 9.2 with those obtained with the 3 df linear element using 10 nodes, the 6 df quadratic element (using 25 nodes) and the exact solution (Chung, 1979).

Table 9.2. Results for the problem of Figure 9.6.

Element	u_7	u_8	u_9
3 df	-	1.232	-
6 df	1.756	1.930	1.826
9 df	2.506	1.834	1.757
Exact	2.509	1.884	1.755

The cubic element models the velocity profile above the crest of the cylinder well because the velocities are nodal freedoms, not the results of approximate 'stress type' calculations.

The solutions for ϕ at the inlet are 3.93 - 4.00, of the expected magnitude.

Therefore the simple cubic formulation is useful for accurate modeling of potential flow and other 'potential type' problems,

9.7. Infinite boundary modeling

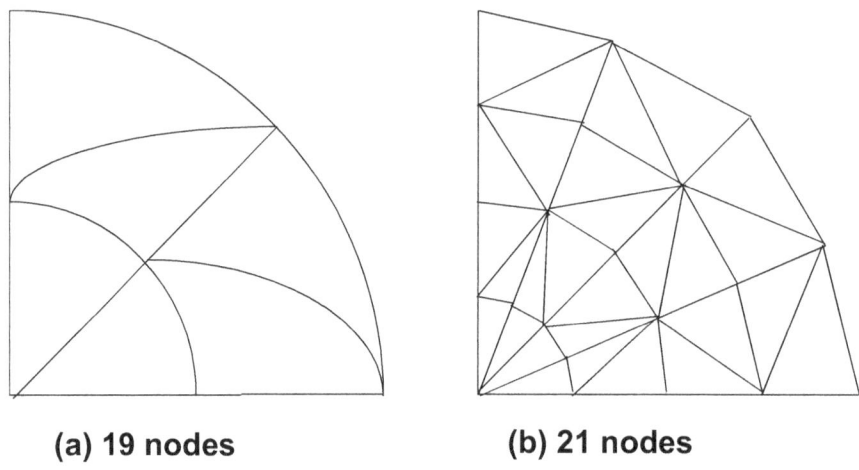

(a) 19 nodes   **(b) 21 nodes**

Figure 9.7. Quadrant of an infinite domain with a point source at its centre.

Figure 9.7 shows a quadrant of a circular domain (of radius 4) with a point source at its centre modeled using (a) 6 node isoparametric elements, and (b) the cubic element of Section 9.3.

To simulate an infinite domain no boundary conditions are imposed other than $\phi = 1000$ at the centre. However, in the fashion of the elastic boundary conditions used by Mohr and Power (1978) for plane thermoelasticity problems, 'stiffnesses' equal to the angle (in radians) subtended by each node's 'share' of the boundary are added to the pivot for each boundary node's ϕ freedom before the final solution of the problem. In Figure 9.7(b), for example, the added values (in degrees) are 11.25 for the two nodes at x = 0 and y = 0 and 22.5 at the other three nodes of the circular boundary.

Table 9.3. Results for the problem of Figure 9.7.

Case	r = 1	r = 2	r = 3	r = 4
6 node FE (1a)	525	367	276	210
6 node FE (1b)	520	365	275	209
Cubic FE (2a)	620	447	341	261
Cubic FE (2b)	520	374	284	218
Expected (3)	520	410	345	299
Expected (4)	480	370	305	259

Table 9.3 shows the results obtained. The expected results are obtained by fitting the appropriate decay function which is (Lamb, 1924)

$$\phi = -(1/2\pi)\phi_0 \ln(r) + C \tag{9.39}$$

and this is that much used in the boundary element method (Brebbia, 1980).

In case (3) (results row 5) the constant C is calculated by substituting $\phi = 520$ at $r = 1$ in Equation 9.39 and using the result to calculate ϕ for the other radii. For case (4) $\phi = 370$ and $r = 2$ is used to obtain row 6.

Cases (1a) and (2a) are the FEM results with only *conditions of infinity* on the boundary. Here a 'natural' decay rate occurs in the meshes used. However in general the desired rate of decay should be modeled by choosing an appropriate C value and setting a corresponding ϕ value as a boundary condition at an inner radius.

This is done for cases (1b) and (2b), setting $\phi = 520$ for the nodes at $r = 1$. The agreement of these two FEM results is now good and agreement with the expected (logarithmic) decay results is reasonable with such coarse meshes and such a rapid decay rate.

Finally, note that from Equation 9.39 it follows that $\partial\phi/\partial r = -\phi_0/(2\pi r)$. This can be set as a boundary condition in Figure 9.6 (at $r = 4$) but gives little change in the results of Table 9.3 (for the cubic element).

9.8. Conclusions

The accurate cubic element used here is easy to derive and is very useful. The simple infinity conditions are perhaps even more important and have wide application, for example to modeling traffic flow (Mohr, 2003) in a similar fashion to the use of FEM to model distribution problems (Mohr, 1999, 2000).

Note that more accurate infinity conditions can be obtained by integrating over the element boundary as in forming the 'contact stiffness matrix' for elastically restrained boundaries (Mohr, 1980).

9.9. References

Argyris JH, Dunne PC, The finite element method applied to fluid mechanics, *Proc. Conf. Computer Methods and Problems in Aeronautical Fluid Dynamics*, University of Manchester 1974, Academic Press, London 1976.

Brebbia CA, *The Boundary Element Method for Engineers*, Pentech Press, Plymouth 1980.

Chung, *Finite Element Analysis in Fluid Dynamics*, McGraw-Hill, New York 1979.

Lamb H, *Hydrodynamics*, 5th edn, Cambridge University Press, Cambridge, 1924.

Martin HC, Finite element analysis of fluid flows, *Proc. 2nd Conf. Matrix Methods in Structural Mechanics*, Wright-Patterson Air Force Base, Ohio 1968.

Mohr GA, Power AS, Elastic boundary conditions for finite elements of infinite and semi-infinite media, *Proc. Instn Civil Engineerrs (London)*, part 2, 65 (1978) 675.

Mohr GA, The finite element contact stiffness matrix for problems involving external elastic restraint, *Computers & Structures* 12 (1980) 79.

Mohr GA, Medland IC, On convergence of displacement finite elements, with an application to singularity problems, *Engineering Fracture Mechanics* 17 (1983) 481.

Mohr GA, *Finite Elements for Solids, Fluids, and Optimization*, Oxford University Press, Oxford, 1992.

Mohr GA, Finite element modeling of distribution problems, *Applied Mathematics & Computation* 105 (1999) 69.

Mohr GA, Optimization of primal and dual network models of distribution, *Computer Methods in Applied Mechanics & Engineering* 188 (2000) 135.

Mohr GA, Power AS, Natural cubic element formulation and infinite domain modelling for potential flow problems, *Australian & New Zealand Institute of Applied Mathematics* (ANZIAM) *Journal,* 44 (2003b) 133-143.

Mohr GA, Finite Element Modelling and optimization of traffic flow networks, *Transportmetrica* v1, n2 (2005) 151-160.

Chapter 10

VISCOUS FLUID FLOW

10.1. Incompressible viscous flow without inertia

For viscous flows the governing differential equations are the Navier-Stokes equations which for an isoptropic two dimensional medium are

$$\varrho(Du/Dt) = X - \partial p/\partial x + \mu(\partial^2 u/\partial x^2) + \mu(\partial^2 u/\partial y^2)$$
$$\varrho(Dv/Dt) = Y - \partial p/\partial y + \mu(\partial^2 v/\partial x^2) + \mu(\partial^2 v/\partial y^2)$$

(10.1)

where
$$Du/Dt = \partial u/\partial t + u(\partial u/\partial x) + v(\partial u/\partial y)$$
$$Dv/Dt = \partial v/\partial t + u(\partial v/\partial x) + v(\partial v/\partial y)$$

(10.2)

are the *total derivatives* of the velocities u,v , the last two terms on the right sides of Equations (10.2) being the *convective inertia* terms, p is the pressure, and ρ and μ are the density and viscosity of the fluid.

Slow incompressible viscous flow in which the inertia terms are neglected is generally referred to as *Stokes flow*. This occurs at very low values of the dimensionless *Reynolds number* Re (usually Re < 1), which appears when the Navier-Stokes equations are non-dimensionalized, practical examples of Stokes flow being in lubrication, visco-elasticity and plasticity problems. The Reynolds number is defined as the ratio of the dynamic pressure ρu^2 to the magnitude of a typical shearing stress $\mu u/L$.

Then the equations for Stokes flow are obtained simply by omitting the left sides of Equations 10.1

$$0 = X - \partial p/\partial x + \mu(\partial^2 u/\partial x^2) + \mu(\partial^2 u/\partial y_2)$$
$$0 = Y - \partial p/\partial y + \mu(\partial^2 v/\partial x^2) + \mu(\partial^2 v/\partial y_2)$$

(10.3)

and these must be coupled with the continuity condition

$$\partial u/\partial x + \partial v/\partial y = 0$$

(10.4)

Numerous finite element formulations for these equations have been developed (Sani et al., 1981, Reddy, 1982), most using Lagrange multipliers or penalty factors to enforce the continuity condition as a *side constraint.*

Perhaps the simplest of these is that developed by Mohr (1984). In this, first introducing the interpolations for u, v, Galerkin weighting is applied to Equations 10.3 and the velocity terms are integrated by parts giving

$$\mu \iint [(\{f_x\}\{f_x\}' + \{f_y\}\{f_y\}')\{u\} + \{f\}\{f_x\}'\{p\}]dxdy = \iint\{f\}'Xdxdy + \mu \int\{f\}'(\partial u/\partial n)dS$$

$$\mu \iint [(\{f_x\}\{f_x\}' + \{f_y\}\{f_y\}')\{v\} + \{f\}\{f_y\}']p\}]dxdy = \iint\{f\}'Ydxdy + \mu \int\{f\}'(\partial v/\partial n)dS$$

(10.5)

where $\{f_x\} = \partial\{f\}/\partial x, \{f_y\} = \partial\{f\}/\partial y.$

Treating Equation 10.4 in like fashion but integrating by parts the first term with respect to x and the second term with respect to y gives

$$\iint(\{f_x\}\{f\}'dxdy)\{u\} + \iint(\{f_y\}\{f\}'dxdy)\{v\} = \int\{f\}[(\partial u/\partial n) + (\partial v/\partial n)]dS = \int\{f\}V_n dS$$

(10.6)

Writing Equations 10.5 and 10.6 in matrix form it is found that a symmetric element matrix is obtained without using negative pressure freedoms or reversing the signs of the continuity constraint equations as in the original formulation (Mohr, 1984):

$$\iint \begin{bmatrix} C & O & A \\ O & C & B \\ A^t & B^t & \alpha I \end{bmatrix} dxdy \begin{Bmatrix} \{u\} \\ \{v\} \\ \{p\} \end{Bmatrix} = \iint \begin{Bmatrix} \{f\}X \\ \{f\}Y \\ \{0\} \end{Bmatrix} dxdy + \int \begin{Bmatrix} \mu\{f\}(\partial u/\partial n \\ \mu\{f\}(\partial v/\partial n \\ \{f\}V_n \end{Bmatrix} dS$$

(10.7)

where

$$A = \{f\}\{f_x\}', \quad B = \{f\}\{f_y\}', \quad C = \mu[\{f_x\}\{f_x\}' + \{f_y\}\{f_y\}']$$

(10.8)

V_n is the outward normal velocity at the domain boundary and O, I are null and unit matrices.

The diagonal entries α for the pressure columns use an arbitrary *small penalty factor* given by

$$\alpha = [\Sigma_i \ \Sigma_j \ C_{ij}^2]^{1/2} \times 10^{-4}/36$$

(10.9)

This value was used for 7 digit computation and with higher precision computation smaller values can be used, for example using 10^{-5} in Equation 10.9 for 8 digit computation with MegaBasic (Mohr, 1992).

In Equation 10.7 the contributions from integration of the continuity equation take the form of Lagrange multiplier constraints which usually have zeroes in the p columns.

In Mohr's formulation the small penalty factors avoid the need for pivoting in the solution routine and might be referred to as inverse penalty factors (Mohr, 1999).

Finally, note that the formulation of Equation 10.7 can easily be converted to an equivalent penalty factor formulation (Mohr, 1992) and the pressures calculated from a modified form of the continuity condition after the solutions for the velocities have been obtained. Retaining the pressure freedoms, however, allows their use for boundary conditions and thus more general modeling.

Note that most formulations for this problem suffer *chequerboard syndrome* (Sani et al., 1982) when the same interpolations are used for both the velocities and pressure. The simplest remedy is, when quadratic interpolation is used for example, to use only linear interpolation for the pressures or, approximately equivalent, use *reduced integration* for the pressure terms in the element matrix.

10.2. Numerical results

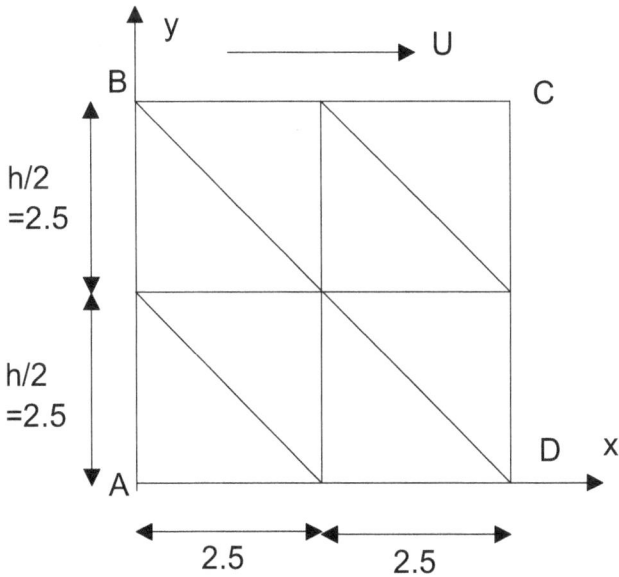

Figure 10.1. Couette flow ($\mu = 1$): boundary conditions are
AB: $p = 4$, $v = 0$, BC: $u = 1$, $v = 0$, CD: $p = 2$, $v = 0$, DA: $u = 0$, $v = 0$

The formulation of Equation 10.7 is easily implemented using six-node triangular elements for which { f } is given by Equations 3.13 and { f_x } and { f_y } are obtained as in Equations 3.34 - 3.35.

Figure 10.1 shows the mesh and boundary conditions used for the Couette flow problem, that is flow between two parallel infinite plates, the upper moving with constant velocity $U = 1$, and under a constant pressure gradient.

Here full quadratic interpolation for the pressures and three point integration at the midside nodes is used, this being a reduced integration for the terms A and B in Equation 10.7.

Table 10.1 compares the solution for the transverse velocity profile obtained with the exact solution (Yuan, 1967)

$$u/U = y/h - (\partial p/\partial x)(hy/2\mu U)(1 - y/h)$$

and Table 10.1 also compares the pressure solutions with the expected linear pressure gradient, indicating good agreement. Here 7 digit computation is used. With 8 digit computation (in MegaBasic) the solutions for the pressure are 4.0, 3.5003, 3.0025, 2.5004, 2.0, a similar result.

Table 10.1. FEM solutions for the problem of Figure 10.1.

Horizontal velocity, u			Pressure, p		
x,y	FEM	Exact	x,y	FEM	exact
2,0	0.0000	0.0000	0,2,5	4.0000	4.0000
2,1	1.1875	1.1875	1.25,2.5	3.4975	3.5000
2,2	1.7500	1.7500	2.5,2.5	3.0012	3.0000
2,3	1.6875	1.6875	3.75.2.5	2.5009	2.5000
2,4	1.0000	1.0000	5,2.5	2.0000	2.0000

For the special case of plane Poiseille flow where $U = 0$ on the upper boundary equally good results are obtained and good results are also obtained for an entry flow problem (Mohr, 1992).

The formulation is easily generalized to deal with axisymmetric flows in pipes by replacing $\partial(\ )/\partial y$ by $\partial(\ )/\partial r$ and $\varpi_i\Delta$ by $2\pi r\varpi_i\Delta$ (where ω_i are the weights) in the numerical integration for the viscosity and other matrices.

10.3. Program for uvp flow

The following BASIC (VB5/6) program uses the six-node quadratic triangular element based on Equation 10.7 for the analysis of Stokes flow. Automatic mesh generation is used and the solution routine deals with the non-zero boundary conditions now required. The data input requirements are described in Table 10.2.

Table 10.2. Data input.

Lines	Data
1	# nodes (NP), elements (NE), b.c. nodes (NB), property sets (NT), half band width (NBW) = Max (node # diff. + 1) in element
1	PENF, POW, giving the multiplier in Equation 10.9 as PENF $\times 10^{POW}$
NT	(PROP(N,1-5)) μ , element thickness, density ρ, body forces X and Y. Here only one data set (N=1) is used (line 290).
1	# nodes in X and Y directions, domain size in X and Y directions
NB	b.c. node # (N), zero/one (one for a b.c.) flags for freedoms u,v,p (IBC(N,I), I=1,2,3) + specified values for u,v,p (SPEC(N,I), I=1,2,3)
variable	node number (NQ) and loads corresponding to freedoms u,v,p - data terminated by a row with NQ=0

The first subroutine MAIN reads data from the file "\flowop\flowop.txt" as described in Table 10.2.

Lines 120-130 set the integration point coordinates and weights and lines 180-320 generate the nodal coordinates and element node number sets for the regular mesh type shown in Figure 10.1. Finally any nodal loads are read and placed in the system load vector, the element routine ELMATF is called for each element and the solution routine SOLVEF is called to assemble and solve the system equations.

```
Attribute VB_Name = "Module1"
DefSng A-H, O-Z: DefInt I-N
Public NP, NE, NB, NT, NBW, NCN, NDF, PENF
Public IBC(100, 3), Q(300), SPEC(100, 3), IMAT(100), CORD(100, 2)
Public PROP(20, 5), NOP(100, 6), CI(6, 2), WF(6)
Public op As Object
Sub main()
rt = Time
```

```
Set op = Form1: Form1.Show
Open "\flowop\flowop.txt" For Input As #1
Open "\flowop\EMATS" For Output As #8
110 NDF = 3: NCN = 6: Dim r(3), IBN(3), SPE(3)
120 CI(1, 1) = 0.5: CI(1, 2) = 0.5: CI(2, 1) = 0: CI(2, 2) = 0.5: CI(3, 1) = 0.5: CI(3, 2) = 0
130 WF(1) = 1 / 6: WF(2) = 1 / 6: WF(3) = 1 / 6
Rem INTEGRATION AT MIDSIDE NODES
150 Input #1, NP, NE, NB, NT, NBW: NBW = NBW - NDF: Debug.Print "NT=", NT
155 Input #1, PENF, POW: PENF = PENF * 10 ^ POW: Rem PENALTY FACTOR
160 For n = 1 To NT: For i = 1 To 5: Input #1, PROP(n, i): Next: Next
170 Input #1, NX, NY, XLIM, YLIM: Rem NX,NY ARE NO. NODES IN X,Y DIRECTIONS
180 NEX = NX - 1: NEY = NY - 1: RNX = NEX: RNY = NE :Rem XLIM,YLIM ARE
190 DX = XLIM / RNX: DY = YLIM / RNY: Rem DOMAIN SIZES IN THOSE DIRECTIONS
200 For i = 1 To NX: For j = 1 To NY
210 RNDX = i - 1: RNDY = j - 1: NN = NY * (i - 1) + j
220 CORD(NN, 1) = RNDX * DX: CORD(NN, 2) = RNDY * DY: Next: Next
Rem Node coords
230 NEX = (NX - 1) / 2: NEY = (NY - 1) / 2
240 For i = 1 To NEX: For j = 1 To NEY
250 NI = (i - 1) * 2 * NY + (j - 1) * 2 + 1: NJ = NI + 2 * NY
260 NS = NEY * (i - 1) + j: NN = 2 * NS - 1
270 NOP(NN, 1) = NI: NOP(NN, 2) = NJ: NOP(NN, 3) = NI + 2
280 NOP(NN, 4) = NI + NY: NOP(NN, 5) = NOP(NN, 4) + 1: NOP(NN, 6) = NI + 1
290 IMAT(NN) = 1: NN = 2 * NS
300 NOP(NN, 1) = NI + 2: NOP(NN, 2) = NJ: NOP(NN, 3) = NJ + 2
310 NOP(NN, 4) = NOP(NN - 1, 5): NOP(NN, 5) = NJ + 1
NOP(NN, 6) = NOP(NN - 1, 4) + 2
320 IMAT(NN) = 1: Next: Next: Rem ONLY ONE PROPERTY SET USED
325 For i = 1 To NB
330 Input #1, n, IBN(1), IBN(2), IBN(3), SPE(1), SPE(2), SPE(3)
For j = 1 To 3: IBC(n, j) = IBN(j): SPEC(n, j) = SPE(j): Next
335 Next
Rem IBC ARE B.C. FLAGS AND SPEC ARE SPECIFIED BOUNDARY VALUES
340 Input #1, NQ, r(1), r(2), r(3): Rem READ NODAL FORCING TERMS
350 If NQ = 0 Then GoTo 380
360 For K = 1 To NDF: IC = (NQ - 1) * NDF + K: Q(IC) = Q(IC) + r(K): Next
370 GoTo 340
380 Rem
390 For n = 1 To NE
400 Call elmatf(n): Next
Close: Open "\flowop\emats" For Input As #8
415 Call solvef
rt = 24 * 3600# * (Time - rt): op.Print "Time = "; rt
430 End Sub
```

The element routine forms the element matrices according to the left side of Equation 10.7 using the quadratic interpolation of Equations 3.13.

Lines 70-90 calculate the projections of the element sides on the axes and then the element area. Lines 130-140 fill the simple 3×2 transformation matrix of Equation 3.35, line 160 'picks up' the integration point coordinates (at the midside nodes) and lines 200-220 fill the matrix of Equation 3.34.

Lines 230-240 form the matrix G of Equation 3.36 and in lines 260-330 it is deployed in the element matrix according to Equation 10.7, line 370 including the small penalty factor terms of Equation 10.9.

Lines 390-490 use a simple basis transformation (Meek, 1971; Mohr, 1992) to suppress the midside pressure freedoms but are not used here.

Finally line 510 writes the element matrices to a file from which they will be used by the solution routine.

```
Attribute VB_Name = "Module2"
DefSng A-H, O-Z: DefInt I-N
Sub elmatf(n)
30 Dim S(18, 18), F(6), TEMP(3, 6), Z(2, 3), V(3, 6), T(18, 18), TT(18, 18)
40 L = IMAT(n): VISC = PROP(L, 1): TH = PROP(L, 2): DENS = PROP(L, 3)
50 FX = PROP(L, 4): FY = PROP(L, 5): Rem COLLECT ELEMENT PROPERTIES
60 i = NOP(n, 1): j = NOP(n, 2): K = NOP(n, 3)
70 X21 =CORD(j,1) -CORD(i,1): X32 =CORD(K,1) -CORD(j,1)
X13 =CORD(i,1) -CORD(K, 1)
80 Y21 =CORD(j 2) -CORD(i,2): Y32 =CORD(K,2) -CORD(j,2)
Y13 =CORD(i,2) -CORD(K, 2)
90 A = X21 * Y32 - X32 * Y21: Rem A=TWICE ELEMENT AREA
100 For i = 1 To 3: NI = i + 3: NF = NOP(n, NI) * NDF - 2
110 Q(NF + 1) = Q(NF + 1) + A * FY / 6: Rem ADD BODY FORCES TO LOAD VECTOR
120 Q(NF) = Q(NF) + A * FX / 6: Next
130 Z(1, 1) = -Y32 / A: Z(1, 2) = -Y13 / A: Z(1, 3) = -Y21 / A
140 Z(2, 1) = X32 / A: Z(2, 2) = X13 / A: Z(2, 3) = X21 / A
150 For II = 1 To 3: Rem COMMENCE INTEGRATION LOOP ##################
160 F1 = 4 * CI(II, 1): F2 = 4 * CI(II, 2): F3 = 4 - F1 - F2
170 C1 = CI(II, 1): C2 = CI(II, 2): CC = C1 + C2
180 F(1) = 2 * C1 * C1 - C1: F(2) = 2 * C2 * C2 - C2: F(3) = 1 - 3 * CC + 2 * CC * CC
190 F(4) = 4 * C1 * C2: F(5) = 4 * C2 * (1 - CC): F(6) = 4 * C1 * (1 - CC): Rem Eqns 3.13
200 V(1, 1) = F1 - 1: V(1, 4) = F2: V(1, 6) = F3
210 V(2, 2) = F2 - 1: V(2, 4) = F1: V(2, 5) = F3: Rem Eqn 3.39
220 V(3, 3) = F3 - 1: V(3, 5) = F2: V(3, 6) = F1
230 For i = 1 To 2: For j = 1 To 6: TEMP(i, j) = 0: For K = 1 To 3
240 TEMP(i, j) = TEMP(i, j) + Z(i, K) * V(K, j): Next: Next: Next
250 VOL = TH * A * WF(II): Sum = 0
260 For i = 1 To 6: For j = 1 To 6
270 G = VOL * VISC * (TEMP(1, i) * TEMP(1, j) + TEMP(2, i) * TEMP(2, j))
280 S(3 * i - 2, 3 * j - 2) = S(3 * i - 2, 3 * j - 2) + G: Rem FILL ELEMENT MATRIX
290 S(3 * i - 1, 3 * j - 1) = S(3 * i - 1, 3 * j - 1) + G
300 S(3 * i - 2, 3 * j) = S(3 * i - 2, 3 * j) + VOL * F(i) * TEMP(1, j)
```

```
310 S(3 * i - 1, 3 * j) = S(3 * i - 1, 3 * j) + VOL * F(i) * TEMP(2, j)
320 S(3 * i, 3 * j - 2) = S(3 * i, 3 * j - 2) + VOL * TEMP(1, i) * F(j)
330 S(3 * i, 3 * j - 1) = S(3 * i, 3 * j - 1) + VOL * TEMP(2, i) * F(j)
340 Sum = Sum + G * G
350 Next: Next
360 For K = 1 To 6
370 S(3 * K, 3 * K) = S(3 * K, 3 * K) + Sqr(Sum) * PENF / 36: Next
380 Next II: Rem END INTEGRATION LOOP ############################
385 GoTo 495: Rem SKIP SUPPRESSION OF MIDSIDE Ps
390 For i = 1 To 18: T(i, i) = 1: Next: Rem FOR THE PRESENT PROBLEM
400 T(12, 3) = 0.5: T(12, 6) = 0.5: T(12, 12) = 0
410 T(15, 6) = 0.5: T(15, 9) = 0.5: T(15, 15) = 0
420 T(18, 9) = 0.5: T(18, 3) = 0.5: T(18, 18) = 0: Rem SUPPRESS MIDSIDE NODE
430 For i = 1 To 18: For j = 1 To 18: Rem PRESSURES
440 TT(i, j) = 0: For K = 1 To 18
450 TT(i, j) = TT(i, j) + S(i, K) * T(K, j): Next: Next: Next
460 For i = 1 To 18: For j = 1 To 18
470 S(i, j) = 0: For K = 1 To 18
480 S(i, j) = S(i, j) + T(K, i) * TT(K, j): Next: Next: Next
490 S(12, 12) = 1: S(15, 15) = 1: S(18, 18) = 1
495 Rem
500 For i = 1 To 18: For j = 1 To 18
510 Write #8, S(i, j): Next: Next: Rem FILE THE ELEMENT MATRICES
520 Debug.Print n: Rem PROGRESS REPORT ONLY
600 End Sub
```

The solution routine reads the element matrices, assuming no more than 8 elements at a node (line 110) and deploys them in the banded system matrix (line 230).

The system equations are solved using Gauss reduction (lines 290 - 550) and stored using 'block solution' (Mohr, 1992). Equations with boundary conditions are flagged with NOB=1 in lines 320 - 360 for reversal of sign, enabling their later use to calculate boundary 'reactions' (not used here).

Thus the equations are stored in a block of size SIZ x LB (set in line 60). When the block is full IBUF (reduced) equations are written to file 'stifm' (lines 560 - 630) and then read back (lines 810 - 880) in reverse order during back substitution (lines 680 - 920).

```
Attribute VB_Name = "Module3"
DefSng A-H, O-Z: DefInt I-N
Sub solvef()
Open "\flowop\stifm" For Random As #10 Len = 600
Open "\flowop\vbout" For Output As #7
45 Dim DIS(3, 100): Rem LINES 320-380,780-790 CHANGE TO HANDLE THE
46 Rem              Rem NONZERO BOUNDARY CONDITIONS IN SPEC( , ).
50 Dim ESM(18, 18), SK(120, 90), SKP(90), T(90)
60 iSIZ = 120: LB = 90: ibuf = iSIZ - LB: nblock = 0
```

```
NRW = 0: NTW = NBW + NDF: NLOAD = NP * NDF: NBN = 1
70 L = 0: n = 1: Rem SIZ,LB ARE STIFFNESS BLOCK DEPTH & WIDTH
80 For i = 1 To 18: For j = 1 To 18
90 Input #8, ESM(i, j): Next: Next: Rem READ FIRST k (ELEMENT MATRIX)
100 L = L + 1: Rem COMMENCE NODE BY NODE FORWARD REDUCTION #######
110 For M = 1 To 8
120 If n = (NE + 1) Then GoTo 280
130 For i = 1 To NCN
140 If NOP(n, i) = L Then GoTo 170: Rem CHECK IF NEXT k NEEDED YET
150 Next
160 GoTo 280
170 Rem
180 For i = 1 To NCN: For j = 1 To NCN
190 For IL = 1 To NDF: IE = (i - 1) * NDF + IL: NR = (NOP(n, i) - 1) * NDF + IL
200 NRE = NR - NRW: Rem NRW = NO. ROWS OF K FILED
210 For JL = 1 To NDF: JE = (j - 1) * NDF + JL: NC = (NOP(n, j) - 1) * NDF + JL
220 NCB = NC - NR + 1: If NR > NC Then GoTo 240
230 SK(NRE, NCB) = SK(NRE, NCB) + ESM(IE, JE): Rem ASSEMBLY OF K
240 Next: Next: Next: Next
245 If n = NE Then GoTo 265
250 For i = 1 To 18: For j = 1 To 18
260 Input #8, ESM(i, j): Next: Next: Rem READ NEXT k
265 n = n + 1
270 Next M
280 Rem
290 NDIF = (NP - L + 1) * NDF: If NDIF > NBW Then LIM = NBW + NDF
320 For ID = 1 To NDF
330 LIM = LIM - 1: IP = ID + NDF * (L - 1): IPE = IP - NRW: r = Q(IP): NOB = 0
340 If IBC(L, ID) <> 0 Then NOB = 1
If NOB = 1 Then RS = -r
350 If NOB = 1 Then r = SPEC(L, ID)
360 If NOB = 1 Then GoTo 380
370 XK = 1 / SK(IPE, 1): Q(IP) = XK * r: GoTo 430
380 Rem
385 Debug.Print L, ID: Rem REPORT SOLUTION PROGRESS
420 Q(IP) = RS + SK(IPE, 1) * r: XK = 1: r = -r: Rem Q(IP) = BOUNDARY 'REACTION'
430 Rem
440 For j = 1 To LIM: JA = j + 1: SKP(j) = SK(IPE, JA): Next
Rem STORE 'ROW MULTIPLIERS'
450 NC = LIM + 1
460 For j = 1 To NC: SK(IPE, j) = SK(IPE, j) * XK: Rem DIVIDE ROW BY PIVOT
470 If NOB = 1 Then SK(IPE, j) = -SK(IPE, j)
Next: Rem NEGATE BOUNDARY ROW
480 If (L + ID - NP - NDF) = 0 Then GoTo 660: Rem END TEST
490 For i = 1 To LIM: NR = IP + i: NRE = IPE + i
500 If SKP(i) = 0 Then GoTo 550
If NOB = 1 Then GoTo 530
NC = LIM - i + 1
510 For j = 1 To NC: JP = j + i
520 SK(NRE, j) = SK(NRE, j) - SK(IPE, JP) * SKP(i): Next
```

```
Rem FORWARD REDUCTION
530 JP = i + 1
540 Q(NR) = Q(NR) - SK(IPE, JP) * r: Rem REDUCTION IN LOAD VECTOR
550 Next i
560 If (IPE + NTW) < iSIZ Then GoTo 630: Rem TEST IF STIFFNESS BLOCK FULL
570 If (NLOAD - NRW) <= iSIZ Then GoTo 630
nblock = nblock + 1
580 For i = 1 To ibuf
nrec = (nblock - 1) * ibuf + i
For j = 1 To LB: T(j) = SK(i, j): Next
590 Put #10, nrec, T: Next:  Rem FILE PART OF STIFFNESS BLOCK
nflag = nblock
600 NRW = NRW + ibuf: Rem NRW = NO. ROWS OF K FILED
610 For i = 1 To LB: For j = 1 To LB: IA = i + ibuf
620 SK(i, j) = SK(IA, j): SK(IA, j) = 0: Next: Next: Rem SHIFT REMAINING ROWS UP
630 Rem
640 Next ID
650 GoTo 100: Rem END NODE BY NODE FORWARD REDUCTION ##############
660 Rem
670 NR = NDF * NP: NRE = NR - NRW: DIS(NDF, NP) = Q(NR): Rem Last disp. known
680 Q(NR) = 0: i = NDF: L = NP
690 GoTo 780
700 L = L - 1: Rem LOOP ON NODES FOR BACK SUBSTITUTION
710 i = i - 1: Rem LOOP ON D.F./NODE FOR BACK SUBSTITUTION
720 NR = NDF * (L - 1) + i: NRE = NR - NRW
730 DIS(i, L) = Q(NR): Q(NR) = 0
740 If LIM < (NBW + NDF - 1) Then LIM = LIM + 1
750 For j = 1 To LIM: JA = j + 1
760 LJ = L + Int((j + i - 1) / NDF): K = i + j - (LJ - L) * NDF
770 DIS(i, L) = DIS(i, L) - SK(NRE, JA) * DIS(K, LJ): Next: Rem BACK SUBSTITUTION
780 If IBC(L, i) = 0 Then GoTo 800
790 Q(NR) = DIS(i, L): DIS(i, L) = SPEC(L, i)
800 Rem
810 If (NRE - NTW) > 0 Or NRW = 0 Then GoTo 880
820 For II = 1 To LB: For j = 1 To LB
830 IA = iSIZ - II + 1: IB = LB - II + 1
840 SK(IA, j) = SK(IB, j): Next: Next
850 NRW = NRW - ibuf
860 For II = 1 To ibuf: Rem READ BACK FILED PARTS OF REDUCED K AS NEEDED
nrec = (nblock - 1) * ibuf + II: Get #10, nrec, T
For j = 1 To LB: SK(II, j) = T(j): Next: Next
nblock = nblock - 1
880 Rem
890 If (i + L - 2) = 0 Then GoTo 930: Rem END TEST
900 If i <> 1 Then GoTo 710: Rem END LOOP ON FREEDOMS/NODE
910 i = NDF + 1
920 GoTo 700: Rem END BACKSUB LOOP ON NODES
930 Rem
935 op.Print "SOLUTIONS FOR U,V,P AT EACH NODE"
940 For n = 1 To NP
```

950 op.Print n, DIS(1, n), DIS(2, n), DIS(3, n): Next
960 For n = 1 To NP: Write #7, n, DIS(1, n), DIS(2, n), DIS(3, n): Next
op.Print "# blocks = ", nflag
980 End Sub

The data file for the problem of Figure 10.1 with 25 nodes and 8 elements is, as described in Table 10.2, line 3 setting viscosity and element thickness= 1:

```
25, 8, 16, 1, 33
1, -5
1, 1, 0, 0, 0
5, 5, 5, 5
1, 1, 1, 1, 0, 0, 4
2, 0, 1, 1, 0, 0, 4
3, 0, 1, 1, 0, 0, 4
4, 0, 1, 1, 0, 0, 4
5, 1, 1, 1, 1, 0, 4
6, 1, 1, 0, 0, 0, 0
10, 1, 1, 0, 1, 0, 0
11, 1, 1, 0, 0, 0, 0
15, 1, 1, 0, 1, 0, 0
16, 1, 1, 0, 0, 0, 0
20, 1, 1, 0, 1, 0, 0
21, 1, 1, 1, 0, 0, 2
22, 0, 1, 1, 0, 0, 2
23, 0, 1, 1, 0, 0, 2
24, 0, 1, 1, 0, 0, 2
25, 1, 1, 1, 1, 0, 2
0,0,0,0
```

The results with this data are those of Table 10.1.

10.4. Conclusions

Unlike most, the formulation of Section 10.2 obtains a symmetric element matrix directly without using negative pressure freedoms or reversing the signs of the continuity constraint equations and, with the use of small penalty factors which are equivalent to Lagrange multipliers (Mohr, 1999), pivoting is not required.

Coding the element matrix of Equation 10.7 is remarkably simple, resulting in a very useful program. With the use of a small artificial viscosity the program can be used for potential flow problems, its advantage then being the use of *primitive variables u,v,p* rather than stream or potential functions, so that boundary condition values for the velocities are able to be set directly.

139

The program does not use 'naturals' (except the usual triangular areal coordinates) or basis transformation, except in a very simple way if the midside node pressure freedoms are suppressed using the simple matrix T formed in lines 400-420 of the element routine.

As Chapter 9 deals with the very basic problem of potential flow, albeit with a fairly sophisticated element, however, it is desirable to follow it with a more general FEM application to fluid flow analysis, one in which the objective of current work is to apply the Stokes equations to each side of a triangle with the objective of, at least in a kernel element matrix, eliminating the need for the continuity condition (Mohr and Power, 2003).

10.5. References

Gresho PM, Sani RL, *Incompressible Flow and the Finite Element Method*, vol. 2: Isothermal laminar flow, Wiley 2000.

Mohr GA, Finite element analysis of viscous fluid flow, *Computers & Fluids* 12 (1984), 217.

Mohr GA, *Finite Elements for Solids, Fluids, and Optimization*, Oxford University Press, Oxford 1992.

Mohr GA, Numerical procedures for input-output analysis, *Applied Mathematics & Computation* 101 (1999) 89.

Mohr GA, Power AS, Natural cubic element formulation and infinite domain modeling for potential flow problems, *J. Aust. and NZ Instn Applied Maths*, 44 (2003) 133-143.

Reddy JN, On penalty function methods in finite element analysis of flow problems, *Int. J. Numerical Methods in Engineering* 2 (1982) 151.

Sani RL, Gresho PM, Leww RL, Griffiths DF, Engleman M, The cause and cure (?) of the spurious pressures generated by certain FEM solutions of the incompressible Navier-Stokes equations: part 2, *Int. J. Numerical Methods in Engineering* 1 (1981) 171.

Chapter 11

OPTIMIZATION OF FEM MODELS

11.1. The method of steepest descent

For multivariate optimization problems the method of steepest descent is one of the most fundamental methods available (Box et al, 1969; Whittle, 1971). This is a *first order gradient method* which is based on allowing a *perturbation* in the Euclidean norm of the variables

$$| \delta x | = \sqrt{(\Sigma \delta x^2)} \qquad (11.1a)$$

resulting in the objective function altering by an amount

$$\delta f = \Sigma (\partial f / \partial x_i)(\delta x_i) + \lambda(\Sigma(\delta x^2 - \Delta^2) \qquad (11.1b)$$

where λ is a Lagrange multipler and Δ is the optimum perturbation, that is, that for which

$$\partial(\delta f) / \partial(\delta x_i) = 0 = \partial f / \partial x_i + 2\lambda (\delta x_i) \quad i = 1 \rightarrow n \qquad (11.2)$$

from which it follows that

$$\{\delta x_i\} = -\{\partial f / \partial x_i\}/2\lambda = -(\text{constant})\{g\} \qquad (11.3)$$

so that the greatest change in the objective function results from search in the direction of the gradient vector $\{ g \}$.

The method is comparable to Newton's method for finding the roots of equations (Mohr, 1992). Indeed for nonlinear optimization of multivariate problems the *second order gradient methods* in which the search direction is given by an equation of the form

$$\{\delta x_i\} = - H^{-1}\{g\} \qquad (11.4)$$

have their mathematical basis in Newton's method.

141

The modern numerical methods, however, begin by assuming $H = I$ (the unit matrix) and gradually form an improved approximation for the Hessian matrix using products of the changes in the gradient vector $\delta\{g\}$.

11.2. Numerical calculation of the gradient vector

In practice numerical methods are usually used to calculate the gradient vector approximately, perturbing each variable in turn by an amount δx_i and noting the change in the objective function δf_i. Then the gradient vector is estimated by the first order finite difference approximations

$$\{g\} = \{\delta f_i/\delta x_i\} \tag{11.5}$$

this being a simple example of the 'vector search methods', many of which use a combination of $\{g\}$ and the vector normal to it as a search direction.

As a simple example of the steepest descent method suppose we wish to minimize the function

$$f = (x_1 - 2)^2 + (x_2 - 1)^2 \tag{11.6}$$

The minimum is very obvious but assuming otherwise the steepest descent search procedure is written using a step length S as

$$\begin{Bmatrix} x_1 \\ x_2 \end{Bmatrix} = \begin{Bmatrix} x_1 \\ x_2 \end{Bmatrix}_{n-1} - S \begin{Bmatrix} \partial f/x_1 = 2x_1 - 4 \\ \partial f/\partial x_2 = 2x_2 - 2 \end{Bmatrix} \tag{11.7}$$

where here the required partial derivatives are known explicitly (but in general would be calculated by the perturbation process of Equation 11.5).

Then beginning at the point (3,3) with $S = 0.2$ Equation 11.7 becomes:

$$\begin{Bmatrix} x_1 \\ x_2 \end{Bmatrix} = \begin{Bmatrix} 3 \\ 3 \end{Bmatrix} - 0.2 \begin{Bmatrix} 2 \\ 4 \end{Bmatrix} = \begin{Bmatrix} 2.6 \\ 2.2 \end{Bmatrix} \tag{11.8}$$

yielding $f = 1.80$. Continuing with gradually increased step lengths the results shown in Table 11.1 are obtained.

Here with $s = 0.6$ we might suspect something but we proceed with $s = 0.8$ just to make sure. Now it is clear that a turning point has been passed, if not earlier, and bisection is used with $S = 0.5$ which, in this simple case, yields the correct solution.

Table 11.1. Steepest descent example.

S	x_1	x_2	f
0	3	3	5
0.2	2.6	2.2	1.8
0.4	2.2	1.4	0.2
0.6	1.8	0.6	0.2
0.8	1.4	-0.2	1.8
0.5	2	1	0

11.3. Constrained nonlinear problems

Constrained nonlinear optimization problems are sometimes solved by stepwise application of the Simplex Method but this is a relatively tedious process. In the present section we describe the SUMT or *Sequence of Unconstrained Minima Technique* (Fiacco & McCormick, 1964) in which constraints are factored by *penalty factors* and added to the objective function. Then search techniques are used with a gradually increased value of the penalty factor to locate the optimum solution with increasing accuracy.

Exterior point methods

In these calculations begin from an exterior point (from the feasible region) and we seek to minimize the function

$$F(x) = f(x) + \beta \Sigma \mid c_i(x) \mid^2 + \beta \Sigma [e_j(x)]^2 \qquad \beta \to \infty \qquad (11.9)$$

where $\mid \mid$ denotes a *step function* which is zero when the inequality constraints $c_i(x)$ are not violated, $e_j(\ x\ )$ denotes an equality constraint and β is a penalty factor.

Then the SUMT technique involves solving the problem as though it were unconstrained using, for example, the steepest descent method with a sequence of gradually increasing values of the penalty factor.

If we begin with $\beta = 0$, for example, we would first obtain the unconstrained minimum. Then as $\beta \to \infty$ the constraints would take control of the solution if appropriate (in view of the use of a step function for the inequality constraints).

Interior point methods

With these SUMT is first applied to the task of minimizing

$$C = \Sigma \, s_i + \Sigma \, [e_i(x)]^2 \tag{11.10}$$

where s_i are slack variables associated with the inequality constraints. This finds a feasible point and then we seek to minimize the function

$$F(x) = f(x) + \beta \, \Sigma [1/c_i(x)] + \beta^{-1/2} \, \Sigma \mid e_i(x) \mid^2 \quad \beta \to 0 \tag{11.11}$$

and now the inverted form of the term for the inequality constraints provides a *response surface* which prevents access to infeasible regions.

Many such slight variations of the SUMT approach have been suggested but we shall restrict attention to the basic form of Equation 11.9.

11.4. An example problem and program

Suppose we add two constraints to the problem of Equation 11.6, giving the constrained nonlinear optimization problem

$$\text{Min:} \quad f = (x_1 - 2)^2 + (x_2 - 1)^2 \tag{11.12}$$

subject to the constraints

$$1 - x_1{}^2/4 - x_2 \geq 0 \tag{11.13a}$$
$$x_1 - 2x_2 + 1 = 0 \tag{11.13b}$$

Then the SUMT problem is stated as

$$\text{Min:} \quad F = f(x_1, x_2) + \beta \mid 1 - x_1^2/4 - x_2 \mid^2 + \beta(x_1 - 2x_2 + 1)^2 \quad \beta \to \infty \tag{11.14}$$

and the gradients of this augmented function are given by

$$
\begin{aligned}
g_1 &= 2x_1 - 4 + 2\beta C(-x_1/2) + 2\beta(x_1 - 2x_2 + 1) \\
g_2 &= 2x_2 - 2 + 2\beta C(-1) - 4\beta(x_1 - 2x_2 + 1)
\end{aligned} \tag{11.15}
$$

where

$$C = \mid 1 - x_1^2/4 - x_2 \mid = 0 \text{ if } (1 - x_1^2/4 - x_2) < 0$$

is a step function.

These gradients are then used to obtain solutions by trial search in the direction of steepest descent:

$$\{x\}_n = \{x\}_{n-1} - S\{g_1, g_2\}$$

when typically search progresses in the fashion shown in Figure 11..1.

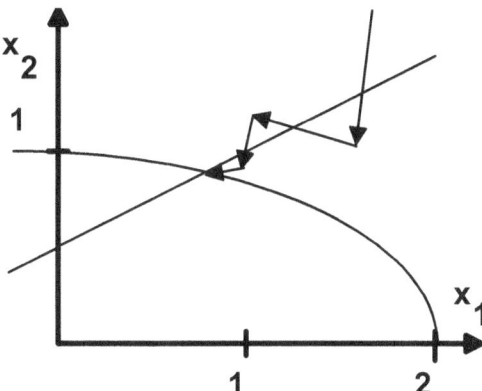

Figure 11.1.
Progress of steepest descent method.

Here the solution is beginning near (2,2), close to the unconstrained optimum at (2,1), and cross-crosses the equality constraint in a relatively inefficient way. This is because the steepest descent direction is influenced by the parabolic constraint and a component tangential to this will be involved in the search direction.

Better results are obtained, however, by using perturbations of each *design variable* in turn (by about 5 or 10%) and calculating approximate gradients from the *finite difference* approximations:

$$g_i = (F_2 - F_1)/\delta x_i \quad \text{for each variable } i$$

This is done in following program and the simple coding can be generalized to deal with many variables and problems in any context.

```
Attribute VB_Name = "Module1"
DefSng A-H, M, O-Z: DefInt I-L, N
Private op As Object: Public X1, X2, B, F
Sub main()
Rem searches: B=1, 0.082 & 0.14; B=100, 0.0047 & 0.00035
Set op = Form1: op.DrawWidth = 3: op.FontItalic = True
Dim C(10, 10): I = 0: M = 1.05: S = 0: op.FontSize = 14: op.FontBold = True
op.PSet (2400, 1400), RGB(255, 255, 255): op.Print "Optimum"
op.PSet (1100, 2100), RGB(255, 255, 255): op.Print ">= constraint"
```

```
op.PSet (700, 500), RGB(255, 255, 255): op.Print "= constraint"
op.PSet (3200, 1000), RGB(255, 255, 255): op.Print "Solution path"
X = 0: Y = 1000: op.Line (X, Y)-(X, Y)
For Z = 0 To 2 Step 0.01
X = Z: Y = 1 - X * X / 4: X = 2000 * X: Y = 3000 - 2000 * Y
op.Line -(X, Y): Next Z
op.Line (0, 2000)-(4000, 0): op.Line (0, 3000)-(6000, 3000): op.Line (0, 3000)-(0, 0)
X1 = 2: X2 = 2: C(1, 1) = X1: C(1, 2) = X2
X = 2000 * X1: Y = 3000 - 2000 * X2: op.Line (X, Y)-(X, Y)
NEWB: a$ = InputBox("B", , , 5000, 4000): B = CSng(a$)
I = I + 1: C(I, 1) = X1: C(I, 2) = X2: S1 = 0: If B = 0 Then GoTo PEND
Call Subb: Debug.Print "IF =", F
F1 = F: X1 = X1 * M: Call Subb: F2 = F: X1 = X1 / M
G1 = (F2 - F1) / (X1 * (M - 1))
X2 = X2 * M: Call Subb: F2 = F: X2 = X2 / M: G2 = (F2 - F1) / (X2 * (M - 1))
NEWS: a$ = InputBox("S", , , 5000, 4000): S = CSng(a$): If S = 0 Then GoTo NEWB
X1 = X1 + (S1 - S) * G1: X2 = X2 + (S1 - S) * G2
Call Subb: Debug.Print X1; " "; X2; "F= "; F
S1 = S
X = 2000 * X1: Y = 3000 - 2000 * X2: op.Line -(X, Y)
GoTo NEWS
PEND: End Sub
Sub Subb()
G = 1 - X1 * X1 / 4 - X2: E = X1 - 2 * X2 + 1: If G > 0 Then G = 0
FU = (X1 - 2) ^ 2 + (X2 - 1) ^ 2
F = FU + B * G * G + B * E * E
End Sub
```

Then by using only two searches with $\beta = 1$ (with searches $S = 0.082$ and $S = 0.14$) and also two searches with $\beta = 100$ ($S = 0.0047$ and $S = 0.00035$) a reasonably good solution can be obtained as

$$x_1 = 0.7321 \text{ and } x_2 = 0.8660 \text{ with } F = 1.6242$$

whereas the exact solution is $x_1 = \sqrt{3} - 1$, $x_2 = \sqrt{3}/2$ and $F = 13.75 - 7\sqrt{3}$.

If exact gradients are used for $\beta = 1, 10, 100, 1000, 10^4, 10^5, 10^6$ six or so searches can be used with each β to obtain a no more accurate result.

This is because using exact gradients is like following tangents on a curve looking for a turning point. The results can be 'miles off'. The approximate finite difference gradients, however, act like a 'chord approximation' and give much better results, the chord (at reasonable intervals) giving a much better idea where a curve is going than the slope

146

11.5. Program for flow optimization

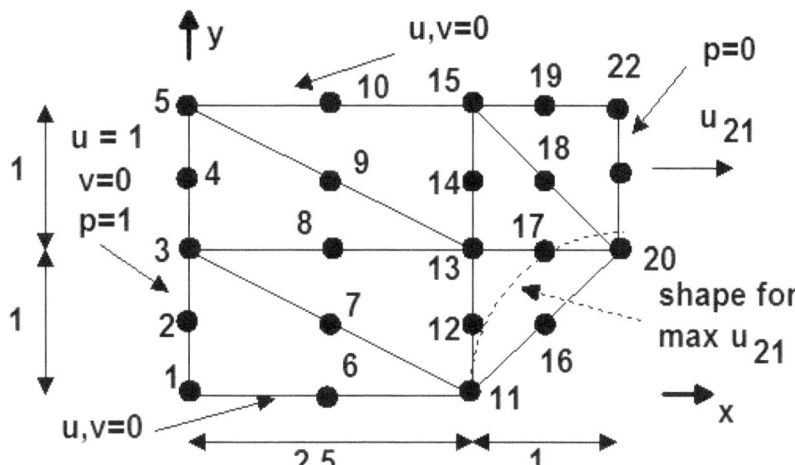

Figure 11.2. Exit flow problem to design a nozzle.

As a further example of optimization using steepest descent the program given in Chapter is 10 modified and applied to the exit flow problem of Figure 11.2. The element formulation closely follows Equation 10.7 and the element routine is coded as described in Section 10.3, except that the isoparametric method of Equation 3.43 (not 3.35) is used to calculate the Cartesian derivatives at each integration point.

In Figure 11.2 vertical movement of nodes 16 and 20 is permitted, 20 per cent increments in these being used to calculate the gradient vector in line 460 of the first subroutine (using 10% increments gave similar results).

Seeking to maximize the exit velocity u_{21} a positive sign is used in Equation 11.4.

At inlet $u, p = 1$ is specified with $p = 0$ at outlet to force the flow. At node 16 the normal velocity should be zero. As skew skew boundary conditions are not available, $v = 0$ is set here as an approximation.

As is often advisable the midside node pressures are suppressed (lines 390-490 of the element routine) to avoid the chequerboard syndrome noted in Chapter 10.

Seeking to approximate inviscid flow $\mu = 0.1$ is used. As the small penalty factor used is 10^{-4} this gives a total 'shift' of 10^{-5}, leaving a minimum of two digits to work with in single precision (7 digit) computation. If a smaller viscosity value is used, therefore, a correspondingly larger small penalty factor should be used. Using $\mu = 0.2$ reduces flow velocities slightly, whilst using $\mu = 0.5$ changes them significantly, and double precision computation gave the same results as single precision.

147

The solution routine is not given here as it is identical to that in Sec. 10.3 except that the statement

Open "\flowop\EMATS" For Input As #8

is added at the beginning and a 'Close' statement at the end.

Table 11.2. Data input.

Lines	Data
1	# nodes (NP), elements (NE), b.c. nodes (NB), property sets (NT), half band width (NBW) = Max (node # diff. + 1) in element NSFV = # of 'shape function' variables
1	PENF, POW, giving the multiplier in Equation 10.8 as PENF $\times 10^{POW}$
NT	(PROP(N,1-5)) μ, element thickness, density ρ, body forces X and Y. Here only one data set (N=1) is used (line 320).
NP	nodal coordinates x = CORD(N,1), y = CORD(N,2)
NE	element node numbers NOP(N,M), M=1 to 6
NB	b.c. node # (N), zero/one (one for a b.c.) flags for freedoms u,v,p (IBC(N,I), I=1,2,3) + specified values for u,v,p (SPEC(N,I), I=1,2,3)
variable	node number (NQ) and loads corresponding to freedoms u,v,p - data terminated by a row with NQ=0
NSFV	'shape function' data: SFCORD(N,1) = node #, SFCORD(N,2) = 1 or 2 for x or y coordinate permitted to change

The data of Table 11.2 is read by the main calling routine which follows.

This also carries out the steepest descent search with simple coding very specific to the example problem of Figure 11.2 but easily generalized. Lines 430-530 form the gradient vector and lines 540-630 conduct the search using trial search lengths input in a message box in line 550. If X = 0 search is ended with the present gradient vector and a new gradient vector is calculated for further search. If X = 99 the program terminates.

```
Attribute VB_Name = "Module1"
DefSng A-H, O-Z: DefInt I-N
Public NP, NE, NB, BT, NBW, NCN, NDF, PENF
Public NBC(100), IBC(100, 3), Q(300), SPEC(100, 3), IMAT(100), CORD(100, 2)
Public PROP(20, 5), NOP(100, 6), CI(6, 2), WF(6), dis(3, 100)
Public op As Object
Sub main()
Set op = Form1: Form1.Show
Open "\flowop\flop.txt" For Input As #1
110 NDF = 3: NCN = 6: XL = 0: Dim r(3), IBN(3), spe(3), SFCORD(10, 2), G(10)
120 CI(1,1) = 0.5: CI(1,2) = 0.5: CI(2,1) = 0: CI(2,2) = 0.5: CI(3,1) = 0.5: CI(3,2) = 0
130 WF(1) = 1 / 6: WF(2) = 1 / 6: WF(3) = 1 / 6: Rem Integration at midsides
```

```
150 Input #1, NP, NE, NB, nt, NBW, nsfv: NBW = NBW - NDF
155 Input #1, PENF, POW: PENF = PENF * 10 ^ POW: Rem Small penalty factor
160 For n = 1 To nt: For i = 1 To 5: Input #1, PROP(n, i): Next: Next
170 For n = 1 To NP: Input #1, CORD(n, 1), CORD(n, 2): Next
180 For n = 1 To NE: For i = 1 To 6
190 Input #1, NOP(n, i): Next: Next
320 For n = 1 To NE: IMAT(n) = 1: Next
325 For i = 1 To NB
330 Input #1, n, IBN(1), IBN(2), IBN(3), spe(1), spe(2), spe(3)
For j = 1 To 3: IBC(n, j) = IBN(j): SPEC(n, j) = spe(j): Next
335 Next: Rem IBC are b.c. flags & SPEC are specified boundary values
340 Input #1, NQ, r(1), r(2), r(3): Rem Read nodal 'loads'
350 If NQ = 0 Then GoTo 380
360 For K = 1 To NDF: IC = (NQ - 1) * NDF + K: Q(IC) = Q(IC) + r(K): Next
370 GoTo 340
380 Rem
385 For n = 1 To nsfv
387 Input #1, SFCORD(n, 1), SFCORD(n, 2): Next: Rem Shape variables
Open "\flowop\EMATS" For Output As #8
390 For n = 1 To NE
400 Call elmatfop(n): Next
415 Call solvef: Rem initial solution
420 Rem
430 For NK = 1 To nsfv: Rem FORM GRADIENT VECTOR
440 VL = dis(1, 21): Rem MERIT FUNC. FOR THIS PROBLEM
450 NS = SFCORD(NK, 1): NXY = SFCORD(NK, 2)
460 DC = CORD(NS, NXY) * 0.2
470 CORD(NS, NXY) = CORD(NS, NXY) + DC: Rem PERTURB SF COORDINATES
Open "\flowop\EMATS" For Output As #8
480 For n = 1 To NE:  Call elmatfop(n): Next
490 Call solvef
500 V = dis(1, 21)
510 DV = V - VL
520 G(NK) = DV / DC: Rem CALCULATE GRADIENT
525 CORD(NS, NXY) = CORD(NS, NXY) - DC: Rem Coords to original values
530 Next NK
540 op.Print G(1), G(2): Rem GRADIENTS (THIS PROBLEM ONLY)
550 XL = X: A$ = InputBox("SL", , , 5000, 4000): X = CSng(A$)
557 If X = 99 Then GoTo 640
560 If X = 0 Then GoTo 420: Rem END SEARCH WITH THIS GRADIENT
570 For NK = 1 To nsfv
580 SN = SFCORD(NK, 1): SXY = SFCORD(NK, 2)
590 CORD(SN, SXY) = CORD(SN, SXY) + G(NK) * (X - XL): Next: Rem SEARCH
Open "\flowop\EMATS" For Output As #8
600 For n = 1 To NE: Debug.Print: Call elmatfop(n): Next
610 Call solvef
620 op.Print dis(1, 21), CORD(16, 2), CORD(20, 2): Rem RESULTS
630 GoTo 550
640 End Sub
```

The element is coded as described in Section 10.3, except that Equation 3.43 (not 3.35) is used to calculate the Cartesian derivatives at each integration point.

```
Attribute VB_Name = "Module2"
DefSng A-H, O-Z: DefInt I-N
Sub elmatfop(n)
10 Rem ELMAT
30 Dim S(18, 18), F(6), TEMP(2, 6), XY(6, 2), DL(2, 6), TJ(2, 2), T(18, 18), TT(18, 18)
40 L = IMAT(n): VISC = PROP(L, 1): TH = PROP(L, 2): DENS = PROP(L, 3)
50 FX = PROP(L, 4): FY = PROP(L, 5): Rem COLLECT ELEMENT PROPERTIES
60 For M = 1 To 6: K = NOP(n, M): XY(M, 1) = CORD(K, 1): XY(M, 2) = CORD(K, 2): Next
70 A = 0
80 For II = 1 To 3: Rem START INTEGRATION LOOP ######################
90 F1 = 4 * CI(II, 1): F2 = 4 * CI(II, 2)
100 DL(1, 1) = F1 - 1: DL(1, 2) = 0: DL(1, 3) = F1 + F2 - 3: Rem EQN 3.41
110 DL(1, 4) = F2: DL(1, 5) = -F2: DL(1, 6) = 4 - 2 * F1 - F2
120 DL(2, 1) = 0: DL(2, 2) = F2 - 1: DL(2, 3) = F1 + F2 - 3
130 DL(2, 4) = F1: DL(2, 5) = 4 - F1 - 2 * F2: DL(2, 6) = -F1
140 For i = 1 To 2: For j = 1 To 2: TJ(i, j) = 0: For K = 1 To 6
150 TJ(i, j) = TJ(i, j) + DL(i, K) * XY(K, j): Rem JACOBIAN, SEE EQN 3.42
160 Next: Next: Next
170 DJ = TJ(1, 1) * TJ(2, 2) - TJ(1, 2) * TJ(2, 1)
180 DD = TJ(1, 1): TJ(1, 1) = TJ(2, 2) / DJ: TJ(2, 2) = DD / DJ: Rem Invert Jacobian
190 TJ(1, 2) = -TJ(1, 2) / DJ: TJ(2, 1) = -TJ(2, 1) / DJ
200 C1 = CI(II, 1): C2 = CI(II, 2): CC = C1 + C2
210 F(1) = 2* C1 * C1 - C1: F(2) = 2* C2 * C2 - C2: F(3) = 1 - 3* CC + 2* CC * CC
220 F(4) = 4 * C1 * C2: F(5) = 4 * C2 * (1 - CC): F(6) = 4 * C1 * (1 - CC)
230 For i = 1 To 2: For j = 1 To 6: TEMP(i, j) = 0: For K = 1 To 2
240 TEMP(i, j) = TEMP(i, j) + TJ(i, K) * DL(K, j): Next: Next: Next: Rem EQN 3.43
250 VOL = TH * WF(II) * Abs(DJ): A = A + VOL
260 For i = 1 To 6: For j = 1 To 6
270 G = VOL * VISC * (TEMP(1, i) * TEMP(1, j) + TEMP(2, i) * TEMP(2, j))
280 S(3 * i - 2, 3 * j - 2) = S(3 * i - 2, 3 * j - 2) + G: Rem AS PER EQN 10.7
290 S(3 * i - 1, 3 * j - 1) = S(3 * i - 1, 3 * j - 1) + G
300 S(3 * i - 2, 3 * j) = S(3 * i - 2, 3 * j) + VOL * F(i) * TEMP(1, j)
310 S(3 * i - 1, 3 * j) = S(3 * i - 1, 3 * j) + VOL * F(i) * TEMP(2, j)
320 S(3 * i, 3 * j - 2) = S(3 * i, 3 * j - 2) + VOL * TEMP(1, i) * F(j)
330 S(3 * i, 3 * j - 1) = S(3 * i, 3 * j - 1) + VOL * TEMP(2, i) * F(j)
340 Sum = Sum + G * G
350 Next: Next
360 For K = 1 To 6
370 S(3 * K, 3 * K) = S(3 * K, 3 * K) + Sqr(Sum) * PENF / 36: Next: Rem Small pf
372 For i = 1 To 6: NF = NOP(n, i) * NDF - 2
374 Q(NF) = Q(NF) + VOL * FX
376 Q(NF + 1) = Q(NF + 1) + VOL * FY: Next
380 Next II: Rem END INTEGRATION LOOP ############################
385 Rem GoTo 495
```

```
390 For i = 1 To 18: T(i, i) = 1: Next
400 T(12, 3) = 0.5: T(12, 6) = 0.5: T(12, 12) = 0
410 T(15, 6) = 0.5: T(15, 9) = 0.5: T(15, 15) = 0
420 T(18, 9) = 0.5: T(18, 3) = 0.5: T(18, 18) = 0: Rem Suppress midside pressures
430 For i = 1 To 18: For j = 1 To 18
440 TT(i, j) = 0: For K = 1 To 18
450 TT(i, j) = TT(i, j) + S(i, K) * T(K, j): Next: Next: Next
460 For i = 1 To 18: For j = 1 To 18
470 S(i, j) = 0: For K = 1 To 18
480 S(i, j) = S(i, j) + T(K, i) * TT(K, j): Next: Next: Next
490 S(12, 12) = 1: S(15, 15) = 1: S(18, 18) = 1
495 Rem
500 For i = 1 To 18: For j = 1 To 18
510 Write #8, S(i, j): Next: Next: Rem FILE THE ELEMENT MATRICES
600 End Sub
```

The data file for the problem of Figure 11.2 is

```
22,7,14,1,33,2
1,-4
0.1,1,0,0,0
0,0, 0,.5, 0,1, 0,1.5, 0,2
1.25,0, 1.25,.5, 1.25,1, 1.25,1.5, 1.25,2
2.5,0, 2.5,.5, 2.5,1, 2.5,1.5, 2.5,2
3,.5, 3,1, 3,1.5, 3,2
3.5,1, 3.5,1.5, 3.5,2
1,11,3,6,7,2
3,11,13,7,12,8
3,13,5,8,9,4
5,13,15,9,14,10
11,20,13,16,17,12
13,20,15,17,18,14
15,20,22,18,21,19
1,1,1,1,1,0,1
2,1,1,1,1,0,1
3,1,1,1,1,0,1
4,1,1,1,1,0,1
5,1,1,1,1,0,1
6,1,1,0,0,0,0
10,1,1,0,0,0,0
11,1,1,0,0,0,0
15,1,1,0,0,0,0
16,0,1,1,0,0,0
19,1,1,0,0,0,0
```

20,0,0,1,0,0,0
21,0,0,0,0,0,0
22,1,1,1,0,0,0
0,0,0,0
16,2
20,2

Table 11.3. Steepest descent results for the problem of Figure 11.2.

	u_{21}	y_{16}	y_{20}
Initial	1.2696	0.5	1.0
Search 1:			
S = 0.5	1.6160	0.9153	1.0653
S = 0.6	1.6477	0.9984	1.0784
S = 0.7	1.6344	1.0814	1.0914
Search 2:			
S = 0.18	1.6634	0.8587	1.2496
S = 0.19	1.6634	0.8509	1.2591
S = 0.2	1.6633	0.8432	1.2687
Search 3:			
S = 0.1	1.6720	0.8994	1.2438
S = 0.2	1.6820	0.9480	1.2284

Table 11.3 shows results for this problem, sufficient results being shown to indicate the simple search procedure used.

In searches 1 and 2 a turning point was located and an approximate maximum for u_{21} found, search ending with this 'maximizing' search length being repeated.

Search 3 began with S = 0.1 and a further search with S = 0.2 gave $y_{16} >$ 0.90 and thus too close to node 17 (for which $y_{17} = 1.0$), so that search was ceased and the results for S = 0.1 taken as the final result for the optimum shape of the outlet in Figure 11.2.

There is no exact solution known for the present problem but the approximate result is in accordance with expectations for subsonic flows (Yuan, 1967) or the 'exponential' shapes of the 'cones' of various sound producing instruments in which, of course, the direction of 'flow' is reversed (Strutt, 1945).

Note, however, that mesh used here is much to coarse to accurately model this flow optimization problem. A symptom of this, no doubt, node 16 has risen close to the level of node 17 (y = 1), deforming this element greatly.

Table 11.4. Steepest descent results for the problem of Figure 11.2.

Search	S	u_{21}	y_{16}	y_{17}	y_{20}
Initial		1.2696	0.5	1.0	1.0
1	0.3	1.5667	0.7492	0.8034	1.2357
2	0.1	1.5708	0.7087	0.8586	1.2593
3	0.2	1.6115	0.7789	0.8074	1.3263
4	0.05	1.6164	0.7482	0.8485	1.3299
5	0.4	1.6499	0.7646	0.8772	1.3925
	0.5	1.6572	0.7686	0.8843	1.4082
	0.6	1.6640	0.7727	0.8916	1.4239
6	0.2	1.6597	0.7985	0.8760	1.4003
	0.3	1.6598	0.8095	0.8718	1.3964
	0.4	1.6591	0.8231	0.8676	1.3924

Table 11.4 shows the results when the y coordinates of nodes 16, 17 and 20 are allowed to move.

In search 5 S = 0.6 moves y_{20} too close to y_{21} (= 1.5) so that search 5 is ceased with S = 0.5 [if search 5 continues to S = 1.4 a maximum u_{21} = 1.6942 is obtained but with y_{20} > 1.5 and thus > y_{21} – an unacceptable situation].

Then the 6th and final search gives an OK turning point with little change to the result of search 5, confirming its validity.

The final result for the horizontal velocity at node 21 is similar to that in Table 11.3, and again nodes 16 and 17 become close.

The overall shape changes and exit velocities obtained in Tables 11.3 and 11.4 are, however, reasonably similar.

The main problem with the foregoing results, however, is the coarseness of the mesh used, requiring excessive shape change for the three elements near the outlet in Figure 11.2. A solution to this difficulty would be the inclusion of 'move limits' in the steepest descent procedure.

Nevertheless, the problem of Figure 11.2 does provide a useful introduction to optimization of FEM flow models using steepest descent.

To confirm the result of Table 11.3 the 'initial solution' (without any shape change via steepest descent search – here obtained using S = 0.00001) with y_{16} = 0.90 and y_{20} = 1.25 (similar to the final result of Table 11.3) gives u_{21} = 1.6781, reasonably similar to the final result for u_{21} in Table 11.3.

With y_{16} = 1 (and thus = y_{17}), however, singularity occurs during computation resulting in overflow and computation ceasing without result, confirming the validity of the restricted search 3 in Table 11.3.

To confirm the result of Table 11.4 the initial solution with y_{16} = 0.8, y_{17} = 0.90, and y_{20} = 1.4 gives u_{21} = 1.6668, in close agreement with the final result of Table 11.4.

11.6. Conclusions

Steepest descent has been widely used for optimization with FEM (Gellatly & Gallagher, 1966; Mohr, 1992, 1994) and the short program of Section 11.4 serves as an introduction to this method.

The program of Section 11.5 then serves as an introduction to its use in a finite element program, albeit very specific to the small example problem studied. Here the constraints of the problem are its boundary conditions and these are imposed in a manner equivalent to the use of penalty factors. Here, of course, these boundary conditions are only equality constraints.

Just as there are almost unlimited applications of FEM, the same is true of such optimization programs. Whilst FEM modeling requires a little care and experiment, however, search for optimum solutions to FEM problems requires a great deal of care indeed.

Though the emphasis of this book is *naturals* and basis transformation it was thought worthwhile to conclude with a chapter on optimization as the elements of earlier chapters can, of course, be used in FEM optimization programs. Note too that suppression of the midside node pressures in lines 390-490 of the element routine is itself a very simple example of basis transformation (Meek, 1971; Mohr, 1992)

Finally, however, note the use of a *natural* argument in forming part of the optimum solution for triangular plates with 'analytical' finite elements with 'deterministic' fields corresponding to isotropic plasticity of the plate (Mohr, 1999).

11.7. References

Box MJ. Davies D, Swann WH, *Nonlinear Optimization Techniques*, ICI Monograph no. 5, Oliver & Boyd, Edinburgh 1969.

Gellatly RA, Gallagher RH, A procedure for automated minimum weight structural design, I: theoretical basis, *Aeronautical Quarterly* 17 (1966) 216.

Fiacco AV, McCormick GP, The sequential unconstrained minimization technique for nonlinear programming, a primal-dual method, *Management Science* 10 (1964) 360.

Meek JL, *Matrix Structural Analysis*, McGraw-Hill Kogashuka, Tokyo 1971.

Mohr GA, *Finite Elements for Solids, Fluids, and Optimization*, Oxford University Press, Oxford 1992.

Mohr GA, Finite element optimization of structures, Computers & Structures 53 (1994) 1217 (part I), 1221 (part II).

Mohr GA, Finite element solutions for optimal triangular plates, *Int. J. Mechanical Sciences* 41 (1999) 1289.

Mohr GA, Optimization of primal and dual network models of distribution, *Computer Methods in Applied Mechanics & Engineering* 188 (2000) 135.

Strutt JW (3rd Baron Rayleigh), *The Theory of Sound*, 2nd edn, Dover, New York 1945.

Whittle P, *Optimization under Constraints*, Wiley, London 1971.

Yuan SW, *Foundations of Fluid Mechanics*, Englewood Cliffs NJ 1967.

11. Optimization of FEM Models

Chapter 12

CONCLUSIONS

12.1. Summary

This short book has tried to take *'naturals'* and *basis transformation* as its main themes, introducing these in Chapters 2 and 3 and following this with new or improved elements developed using these approaches in following chapters.

With the use of natural strains, slopes etc. all elements are triangular. Indeed natural slopes were needed at the outset to establish the standard Hermitian interpolation for a cubic triangular element in Chapter 3.

Thanks to the use of basis transformation the LST (linear strain triangle), or more precisely the six node quadratic Lagrangian triangle, is able to be used to form the basis of most of the key elements derived, namely the QBTP thin plate element (Chapter 4), the DFT elements (Chapter 5), the QBTP + DFT2 facet shell element (Chapter 6), the thick plate element of Chapter 7, the curved shell element of Chapter 8 and the viscous flow element of Chapter 10.

In addition the cubic bases of the BCIZ and CBTP thin plate elements are also able to be used to form elements for potential flow.

These elements are as compact as possible and details of some higher order elements such as quartic and quintic thin plate elements are given by Mohr (1992) where eigenvalue, nonlinear and optimization problems and many other finite element techniques and applications are also discussed.

12.2. Conclusions

It is hoped and expected that the much of the material of this book will prove of wide and continuing interest, for example:

➢ The 9 freedom QBTP plate element. In meshes of practical fineness this is almost as accurate as the very complicated 18 df quintic element.

➢ The CBTP element. This is exactly equivalent to the much used BCIZ element but not as accurate as QBTP. Its advantage over the BCIZ element is that it allows the option of retaining a centroidal w freedom. Both the BCIZ and QBTP elements provide simple examples of the use of natural strains.

➢ The DFT2 element, though it requires a single ϕ_z boundary condition for the constant moment patch test (and should not) it is quite accurate and represents shear stress quite well (whereas the LST does very poorly indeed). Indeed its formulation, requiring only six entries of the form $s_a/8$ in its T matrix to distinguish it from the constant strain triangle, is perhaps fundamental.

➢ The 18 freedom thick plate element (Chapter 7) provides a further example of the application of natural strains, here to transverse shear strains in a plate.

➢ The doubly curved shell element of Chapter 8 incorporates an elegant shell theory in which:

(a) natural strains are used to calculate the curvatures.

(b) natural strains are used to calculate the transverse shear strains.

(c) natural strains are used to calculate the membrane strains.

(d) the natural approach is used to calculate the geometry of the element with only the nodal coordinates as data.

As a result of (d):

(e) natural radial strain components of the membrane strains are able to be calculated.

(f) natural circumferential strain components of the curvatures are able to be calculated.

➢ The accurate cubic element for potential flow is first derived using the simple BCIZ interpolations and then using the standard cubic Hermitian interpolation and the basis transformation of the CBTP element. The simple infinity conditions used to model infinite domains with this element should prove of wide interest..

➢ The viscous flow element of Chapter 10 is remarkably easy to code and, with the use of selective integration by parts of the Stokes and continuity equations the element matrix is symmetric. Then with the use of small penalty factors (equivalent to inverse Lagrange multipliers) pivoting is no longer required.

I hope therefore that the book proves of wide and ongoing interest.

I believe the *natural* approach one that will survive through time. I expect that the 'natural' use of basis transformation along element sides to obtain new elements, for example the accurate QBTP element, will also stand the test of time. **Geoff Mohr**

Appendix A

THE LARGE CURVATURE CORRECTION

A.1. Introduction

As this short book has concentrated a good deal on the use of *natural* coordinates, freedoms and strains in triangular finite elements, together with basis transformations which transform global freedoms and element strains to natural values, it was thought appropriate to append my final paper on the *large curvature correction* (LCC), the initial work having begun in 1979.

As with the numerical differential geometry calculations of Sec. 8.3, where the difficult problem of calculating a twist curvature R_{xy} is avoided, the large natural curvature correction (LCC) could best be incorporated into plate elements using the natural approach as there is no need to determine the form of the correction needed for the twisting curvature, if indeed this is possible.

The LCC correction occurs only when beam slopes are of the order of one radian, an impractical prospect in elastic plate structures, and even then it is numerically elusive. In such applications as military studies of impact failure of steel shells, however, it might be of importance.

The LCC might also be of relevance in Regge calculus studies of the theory of relativity, Regge calculus being the manifestation of FEM in this field.

Geoff Mohr

A.2. The final paper on the LCC

The following paper improves on the two earliest papers of 1980 and 1981 by Mohr and Milner, combining the final two papers of 2001 and then adding further results obtained using double precision computation and extrapolation.

THE LARGE CURVATURE CORRECTION IN MATRIX STRUCTURAL ANALYSIS

G. A. MOHR

Abstract

Finite element analysis of large displacement problems often includes the effect of large slopes upon the extensional strains but ignores their effect in the curvature calculations. In the present work a fictitious load method and two alternative approaches to calculation of this 'large curvature correction' are considered.

Four freedoms per node, rather than the usual three, are found to give improved results and a useful 'two-way' extrapolation technique is used to confirm the validity of the large curvature correction.

More accurate results are then obtained to fully confirm these LCCs and an alternative LCC proposed by Milner is also tested. A quintic/cubic formulation is then tested and cumulative stress calculations are introduced without difficulty. Finally, the accuracy of the curvilinear curvature correction is convincingly demonstrated using large numbers of elements and load steps.

1. Introduction

Large displacements are important in the study of integral geometry problems [6], in earthquake analysis [1], in the analysis of structural failure and collapse [4, 7], and in the study of the long-term durability of structures [15].

Large displacement problems are readily modelled by the finite element method using load stepping in which the load is applied in increments, updating the element stiffness matrices using coordinate transformation [12]. This approach is only approximate, however, and does not model the actual strains developed by large displacements.

For 'moderately' large displacements, therefore, strain calculations should include the effect of the beam slope [12, 17], that is

$$\varepsilon = \partial u / \partial x + (\partial v / \partial x^2) \tag{1.1}$$

When the slopes themselves become significant, however, a corresponding adjustment to calculation of the curvature (a generalized strain) is required.

In Cartesian coordinates this is simply

$$\chi = \partial^2 v / \partial x^2 [1 + (\partial v / \partial x)^2] \qquad (1.2)$$

and this correction is easily included in finite element analysis [10].

In the present paper an error in the Cartesian large curvature correction (LCC) used by Mohr and Milner [10] is corrected.

An alternative approach based on the curvilinear correction proposed by Milner [8] is then developed and found to be of comparable accuracy to the fictitious load method of Kohnke [5] in which transcendental functions are used to provide accurate solutions.

The present paper includes earlier published results with no more than 20 elements and 40 load steps, when extrapolation was used to help assure that the solution was indeed converging correctly [13, 14]. Additional results with up to 100 elements and 100 load steps are also obtained, however, to establish beyond all doubt that the present 'LCC' does indeed converge to the correct solution.

2. Cartesian curvature correction

Considering the six freedom beam element as an example, large displacement analysis commences with a first linear trial solution $\{d\}$ for the element displacements, after which the element curvatures are calculated as the numerator of Eqn (1.2) and the extensional strains from Eqn (1.1). Then the generalized element stresses are given by

$$\sigma = D \begin{Bmatrix} \varepsilon \\ \chi \end{Bmatrix} = \begin{bmatrix} EA & 0 \\ 0 & EI \end{bmatrix} [B_0 + B_L / 2]\{d\} = D\bar{B}\{d\} \qquad (2.1)$$

where, using linear and cubic interpolation with $s = 0\ to\ 1$, we have

$$\bar{B} = \frac{1}{L}\begin{bmatrix} -1 & f_1 v_s / 2 & f_2 v_s / 2 & 1 & f_3 v_s / 2 & f_4 v_s / 2 \\ 0 & g_1 & g_2 & 0 & g_3 & g_4 \end{bmatrix} \qquad (2.2)$$

$$f_1 = 6s^2 - 6s,\ f_2 = L(3s^2 - 4s + 1),\ f_3 = 6s - 6s^2,\ f_4 = L(3s^2 - 2s) \qquad (2.3)$$

$$g_1 = (12s - 6) / L,\ g_2 = 6s - 4,\ g_3 = (6 - 12s) / L,\ g_4 = 6s - 2 \qquad (2.4)$$

$$v_s = \{0, f_1, f_2, 0, f_3, f_4\}^t\, T_0\, \{d\} / L = \partial v / \partial s \qquad (2.5)$$

is the interpolation of the Cartesian slope derivative in the local (at the start of the load step) frame of reference, with T_0 being the standard 6x6 element coordinate transformation matrix applying at the start of the load increment.

APPENDIX A: THE LARGE CURVATURE CORRECTION

Considering a *virtual* variation in the total extensional strain as

$$\delta\varepsilon^* = \delta(\partial u/\partial/x) + \delta(\partial v/\partial x)^2/2 = \delta(\partial u/\partial x) + (\partial v/\partial x)\delta(\partial v/\partial x) \quad (2.6)$$

the element reactions and stiffness matrix are calculated by numerical integration using

$$\{q_r\} = \sum B^{*t} D \overline{B} \{d\}\omega_i L \quad (2.7)$$

$$k = \sum B^{*t} DB^* \omega_i L \quad where \ B^* = B_0 + B_L \quad (2.8)$$

and three point Gauss quadrature is used to exactly integrate Eqns (2.7) and (2.8).

Note that when load stepping is incorporated, as in the present work, both B matrices must be postmultiplied by the coordinate transformation matrix T_0. Note, however, that the element length (L) in Eqns (2.2)-(2.5) and (2.7),(2.8) is the latest value at the current iteration.

Note too that the ESM k can also be calculated using the standard formula for 6 df beam elements to obtain a local ESM k_L, the global ESM being given by

$$k = T^t k_L T \quad (2.9)$$

where T is for the latest element geometry. In the present work, however, the more general numerical integration calculation is used to fully test the corrections introduced but Eqn (2.9) is used for Kohnke's fictitious load method [5] and for simple load stepping.

Now if the curvatures are calculated according to Eqn (1.2) then row 2 of Eqn (2.2) is simply divided by the cube of factor F where

$$F = [1 + (\partial v/\partial x)^2]^{1/2} = [1 + v_x^2] \quad (2.10)$$

where v_x is obtained from Eqn (2.5) with T_0 omitted.

Corresponding to Eqn (2.6) a virtual curvature increment is calculated as

$$\delta\chi^* = \delta v_{xx} F^{-3} + v_{xx}\delta(F^{-3}) = \delta v_{xx} F^{-3} - 3v_{xx}v_x F^{-5}\delta(v_x) = \delta(v_{xx}/F^3)[1 - 3v_x^2/F^2] \quad (2.11)$$

where v_{xx} is the second derivative of v with respect to x and reciprocity is used to 'exchange' incrementation in this for incrementation in v_x in the second term, so that B^* is obtained simply by multiplying row 2 of Eqn (2.2) by the last [ ] term in Eqn (2.11).

Note that in the original work of Mohr and Milner [10] v_x was not squared in this last result, such reciprocity steps not then being well understood.

162

3. Fictitious loads

The fictitious load procedure of Kohnke [5] uses Eqn (2.9) to update the element matrices and corrections to the initial (small displacement) solutions can be obtained by applying the fictitious loads

$$\{q_f\} = \{F_1, -F_2, -F_3, -F_1, F_2, -F_3\} \tag{3.1}$$

where

$$F_1 = EA(1 - \cos\beta),\ F_2 = 12EI(\beta - \sin\beta)/L_0^2,\ F_3 = 6EI(\beta - 4\sin\beta)/L_0 \tag{3.2}$$

and the angle β is the chord slope of the element relative to that at the start of the current load step.

Following Kohnke L_0, the element length at the beginning of the current load step, is used here (and L' the current length elsewhere in this work).

Then the loads F_1 correspond to an implicit extensional strain arising from the rotation β (the first term of which is captured by Eqn (1.1)) and F_2, F_3 correspond to accurate 'arc' calculation of curvature rather than the usual approximation as a second derivative of lateral displacement.

Then the nodal coordinates are iteratively updated using

$$\{X\}_i = \{X\}_{i-1} + \{D\}_i - \{D\}_{i-1} \tag{3.3}$$

for iteration i, noting that at the start of each load step displacements are zero (for that step) and its external loads are applied only then.

4. Curvilinear curvature correction

Results with the 6 df beam element and the interpolated LCCs are too stiff and require many load steps for reasonable accuracy. In the test problem studied here we have a slope of one radian at the end of the beam. This is, indeed, an arch shape and as in shell analysis, linear displacement interpolations are inevitably too stiff in such situations [16]. Obviously, therefore, it is helpful to use a consistent order of interpolation [17].

Then, allowing a fourth nodal freedom, the first derivative of u, and thence using cubic interpolation for both u and v, the matrix of Eqn (2.2) is replaced by

$$\overline{B} = \frac{1}{L}\begin{bmatrix} f_1 G & f_2 G & f_1 v_s/2 & f_2 v_s/2 & f_3 G & f_4 G & f_3 v_s/2 & f_4 v_s/2 \\ 0 & 0 & g_1 & g_2 & 0 & 0 & g_3 & g_4 \end{bmatrix} \tag{4.1}$$

where

$$v_s = \{0, 0, f_1, f_2, 0, 0, f_3, f_4\}^t\, T_0\, \{d\}/L \tag{4.2}$$

is the interpolated value of the 'local' slope (the first derivative of v wrt to s), and the factor G is given by

$$G=(1+u_s)/2 \qquad (4.3)$$

where
$$u_s=\{f_1,f_2,0,0,f_3,f_4,0,0\}^t\, T_0\, \{d\}/L \qquad (4.4)$$

and this corresponds to inclusion (in local coordinates) of the 'Pythagorean complement' of the last term in Eqn (1.1).

Then, in forming matrix B^*, the $1/2$ factors of Eqns (4.1) and (4.3) are removed. Now the curvature is also calculated in terms of the local coordinate as

$$\chi=(\partial^2v/\partial s^2)/F, \quad F=[1-v_s^2]^{1/2} \qquad (4.5)$$

so that row 2 of Eqn (4.1) is divided by the factor F.

Now the virtual increment is given by

$$\delta\chi^*=\delta v_{ss}F^{-1}+v_{ss}\delta(F^{-1})=\delta v_{ss}F^{-1}+v_{ss}v_sF^{-3}\delta(v_s)=(\delta v_{ss}/F)[1+v_s^2/F^2] \qquad (4.6)$$

where v_{ss} is the second derivative of v with respect to s and again reciprocity is used to obtain the final result. Thus in forming matrix B^* row 2 of Eqn (4.1) is divided by the last [ ] factor of Eqn (4.6).

5. Numerical results

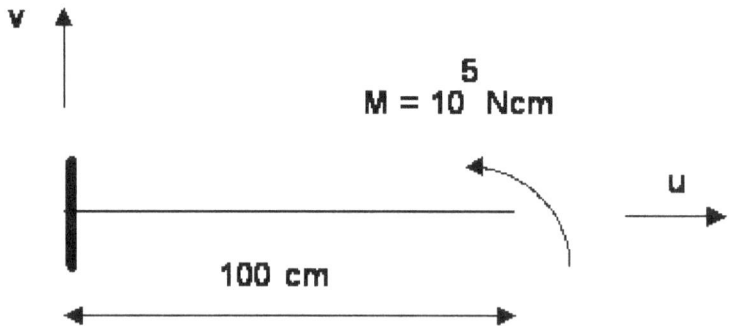

FIGURE. 1. Large displacement of beam, $A = 10$ cm^2, $I = 5$ cm^4, $E = 2 \times 10^6$ cm^2

Figure 1 shows a simple test problem and Table 1 gives the results obtained for the displacements at the free end using both 4 and 10 elements and from 10 to 40 load steps with from 1 to 5 iterations at each load step.

The analytical solution, with $\phi_e = 1$ radian as the rotation at the free end and $R = 100$, is

$$u = R(1-\sin(\phi_e)) = 15.8529, \quad v = R(1-\cos(\phi_e)) = 45.9698 \tag{5.1}$$

TABLE 1.
Solutions using 3 d.f. per node; e = # elements, s = # load steps, i = # iterations

e	s	i	u	v	ϕ
Load stepping					
4	10	1	-14.4982	47.2845	1.0075
10	10	1	-14.6866	47.2181	1.0078
10	20	1	-15.2707	46.6073	1.0040
10	40	1	-15.5499	46.3003	1.0020
With strain correction					
4	10	3	-9.2815	36.6477	0.7408
10	10	3	-13.5917	43.1034	0.9145
With Cartesian curvature correction					
4	10	3	-9.3152	36.6826	0.7441
10	10	3	-13.7880	43.3140	0.9261
10	10	5	-13.7831	43.3070	0.9259
10	20	3	-15.2040	45.2057	0.9780
10	40	3	-15.6558	45.7853	0.9942
With curvilinear curvature correction					
4	10	3	-9.2879	36.6699	0.7401
10	10	3	-13.5845	43.1156	0.9125
10	10	5	-13.5724	43.0090	0.9121
10	20	3	-15.1177	45.1203	0.9730
10	40	3	-15.6305	45.7586	0.9928
With fictitious loads					
4	10	3	-15.6042	46.0627	1.0000
10	10	3	-15.7876	45.9617	1.0000
10	10	5	-15.7876	45.9617	1.0000
10	20	3	-15.8104	45.9819	1.0000
10	40	3	-15.8160	45.9872	1.0000
Fictitious loads without moment correction					
10	10	3	-15.7565	45.9358	1.0000
Exact			-15.8529	45.9698	1.0000

First simple load stepping (and Eqn (2.9)) are used to obtain a crude approximation, but a useful point of reference for results that follow.

Next the correction of Eqn (2.6) is included, using the formulation of Eqns (2.2)-(2.8). The result is too stiff, being further from the exact solution. Including the Cartesian curvature correction of Eqns (2.10) and (2.11), however, results in only a small improvement. Similar results are obtained with the curvilinear correction of Eqns (4.5) and (4.6).

Next results obtained using fictitious loads are given. Note that double precision computation (14 dp) is used for this method for consistency with the original 11 dp precision work [10], whereas 8 dp precision is used elsewhere in the present work. Consequently the result for 10e/10s/3i is exactly that obtained by Mohr and Milner [10] and is very close to the exact solution. Finally, the order of magnitude of the curvature correction is illustrated by setting $I = 0$ in Eqns (3.2), giving the last row of results in Table 1.

As noted in Sec. 3, length L_0 (at start of a load step) is used in Eqns (3.2) but the result with s = 40 is the same if current length L' is used.

Table 2 shows the results obtained with the 8 df element. First load stepping is applied, the results being close to the corresponding ones in Table 1 for the 6 df element. Note that the additional u_x freedoms are not used as boundary conditions and that u_x at the free end is nearly zero in the load stepping solution and is close to 0.05 in the other cases.

Next the extensional strain correction of Eqn (1.1) is included, yielding results close to the exact solution.

Then the additional correction of Eqns (4.3) and (4.4) is also added, yielding only a very small change in results, as expected.

Next the Cartesian curvature correction of Eqns (2.10) and (2.11) is included but many load steps are required to obtain reasonable accuracy. Then the curvilinear correction of Eqns (4.5) and (4.6) is used. Improved results of comparable accuracy to those of the fictitious load method in Table 1 are obtained.

TABLE 2. Solutions using 4 d.f. per node

e	s	i	u	v	ϕ
Load stepping					
4	10	1	-14.4958	47.2806	1.0074
10	10	1	-14.6775	47.2034	1.0076
10	20	1	-15.2633	46.5960	1.0038
10	40	1	-15.5536	46.3056	1/0021
With strain correction					
4	10	3	-15.6782	46.1027	0.9999
10	10	3	-15.8489	46.0171	1.0000
With additional strain correction					
4	10	3	-15.6799	46.1009	0.9999
10	10	3	-15.8507	46.0154	1.0000
With Cartesian curvature correction					
4	10	3	-15.9939	46.4164	1.0173
10	10	3	-16.1707	46.3273	1.0173
10	10	5	-16.6310	47.1153	1.0271
10	20	3	-15.9045	46.0715	1.0042
10	40	3	-15.8397	46.0092	1.0010
With curvilinear curvature correction					
4	10	3	-15.5643	45.9843	0.9932
10	10	3	-15.7333	45.9006	0.9933
10	10	5	-15.7329	45.9002	0.9934
10	20	3	-15.7972	45.9660	0.9983
10	40	3	-15.8130	45.9829	0.9996
Exact			-15.8529	45.9698	1.0000

6. Extrapolation

Table 3 shows the results obtained for the lateral deflection at the free end using 1, 2 and 4 elements and 'complete' (all terms included) fictitious load and curvilinear curvature corrections (respectively with the 6 df and 8 df elements).

The fictitious load result is poor with one step and two elements and with four (and also ten) elements divergence occurs. One remedy is to use a convergence factor [3] of 0.5 but two load steps also avoids difficulty and solutions with two or more elements and steps are reasonable.

TABLE 3. Solutions with one, two and four elements

s	i	$e = 1$	$e = 2$	$e = 4$
Fictitious loads (3 df/node)				
1	8	46.3992	44.6924	44.2942
2	5	47.5265	45.9006	45.5311
4	4	47.8337	46.2994	45.9322
10	3	47.9246	46.4262	46.0627
Extrapolated (two way)			45.9531	45.9665
Curvilinear curvature correction (4 df/node)				
1	8	38.3713	39.1412	39.7264
2	5	43.9941	43.9222	44.0216
4	4	46.4692	45.6840	45.5085
10	3	47.5207	46.2849	45.9843
Extrapolated (two way)			45.9589	45.9668
Exact:			45.9698	

With four steps the curvature correction results are reasonable but for accurate results four elements and 10 steps (or vice versa) are recommended.

Finally h^2 extrapolation [12] is applied in a 'two-way' fashion to the results with 4/10 steps and 1/2 and 2/4 elements, yielding good results for both the fictitious load and curvilinear corrections. For the latter, for example, extrapolating the results for $e = 2$ and $s = 4,10$ we obtain

$X = 46.2849 + (46.2849 - 45.6840)/5.25 = 46.3994$.

For $e = 4$ and $s = 4, 10$ we get

$Y = 45.9843 + (45.9843 - 45.5085)/5.25 = 46.0749$

and extrapolating these results (for $e = 2, 4$) the final result is

$Z = Y + (Y - X)/3 = 45.9668$, as shown in Table 3.

Such 'two-way' extrapolation is itself a useful technique with many applications, most obviously in the field of financial market analysis, for example.

Finally, note that h^2 extrapolation is appropriate for this element was confirmed using trial in the simple equation for this suggested by Mohr [12, Eqn 3.62].

7. Additional results

Table 4 gives the results obtained for the displacements at the free end in the problem of Fig. 1 using 3 df/node and simple load stepping with 20 elements.

Next the results obtained with 20 elements and fictitious loads are given. The final result for v is close to that obtained by Kohnke [5] with 20 elements, namely $v = 45.9746$. Note that double precision computation is used for this method (in line with previous work) but single precision (8 dp) elsewhere in the present work.

168

TABLE 4.
Results with 20 elements, e = # elements, s = # load steps, i = # iterations/step

e	s	i	u	v	ϕ
Load stepping (3 df/node):					
20	10	1	-14.6680	47.1357	1.0067
20	20	1	-15.2533	46.5282	1.0029
20	40	1	-15.5616	46.2651	1.0017
Fictitious loads (3 df/node):					
20	10	3	-15.8137	45.9473	1.0000
20	20	3	-15.8366	45.9676	1.0000
20	40	3	-15.8428	45.9728	1.0000
Load stepping (4 df/node):					
20	10	1	-14.7010	47.1887	1.0076
20	20	1	-15.3197	46.6265	1.0046
20	40	1	-15.5419	46.2367	1.0012
With strain corrections (4 df/node):					
20	10	3	-15.8750	46.0032	1.0000
20	20	3	-15.8519	45.9814	1.0000
20	40	3	-15.8461	45.9762	1.0000
With Cartesian LCC (4 df/node):					
20	10	3	-16.1959	46.3145	1.0173
20	20	3	-15.9300	46.0581	1.0043
20	40	3	-15.8656	45.9953	1.0010
With curvilinear LCC (4 df/node):					
20	10	3	-15.7576	45.8890	0.9934
20	20	3	-15.8224	45.9528	0.9983
20	40	3	-15.8388	45.9690	0.9996
Exact:			-15.8529	45.9698	1.0000

Next the results using load stepping with 4 df/node (cubic interpolation for both u and v) are given. These are much the same as those with 3 df/node.

Then the results with the extensional strain calculated as in Sec. 4, that is as

$$\varepsilon = (\partial u / \partial s)[1 + u_s / 2] + v_s (\partial v / \partial s) / 2 \tag{7.1}$$

are given, followed by those with either the Cartesian or curvilinear LCCs added. The final result for the latter is correct to five figures but note the small difference from the result without any LCC (results row 12).

This shows that the LCC is, indeed, elusive and is more clearly seen to advantage when only a few elements are used, for example in Table 3. But first we must choose the most accurate LCC to use and be sure that it is correct and the present results help in this.

Finally the fictitious load method was tried with 4 df/node but to this point has not yielded satisfactory results.

8. An alternative LCC

Milner [8] proposed an alternative LCC given by

$$\delta\chi *=\delta v_{ss}/F+v_{ss}v_s\delta v_s/F^3 \tag{8.1}$$

with F as in Eqn (4.5), but moment loadings of

$$M/\cos\phi \quad where \quad \partial v/\partial s=v_s=\tan\phi \tag{8.2}$$

were reportedly used to obtain excellent results with only one load step and 1-3 elements.

Table 5 shows results obtained using this method, including an additional column $v/cos\phi$ to try and obtain Milner's claimed highly accurate solutions [8].

TABLE 5. Results with Milner's method (using 4 df/node)

e	s	i	u	v	ϕ	$v/cos\phi$
1	1	6	-8.6064	35.5403	0.6729	44.1689
2	1	6	-9.0070	35.5098	0.6805	45.6864
4	1	6	-8.9838	35.5222	0.6781	45.6118
10	1	10	-8.8795	35.3368	0.6743	45.2618
20	1	10	-8.8852	35.3510	0.6745	45.2618
4	10	3	-15.5034	45.8806	0.9933	84.0368
10	10	3	-15.6722	45.7966	0.9934	83.8982
10	20	3	-15.7812	45.9403	0.9983	84.8059
10	40	3	-15.8092	45.9763	0.9996	85.0378
20	10	3	-15.6961	45.7844	0.9934	83.8777
20	20	3	-15.8072	45.9275	0.9983	84.7855
20	40	3	-15.8343	45.9618	0.9996	85.0097
20	50	3	-15.8383	45.9668	0.9997	85.0406
Exact:			-15.8529	45.9698	1.0000	

Contrary to his claims, with a single load step v is only about 35 but $v/cos\phi$ is in the vicinity of the exact solution for v. This is merely a mistake, however, and neither v nor ϕ are close to the exact solution. Indeed the results for $'v = v/cos\phi'$ are not so close to the exact solution as suggested by Milner's figures (for example, $'v' = 45.968$ for 3e/1s/6i).

The results with 10 or more load steps are satisfactory, however, showing that it is possible to have two incremental quantities on the RHS in Eqn (7.1). The results, for example for 20e/10s/3i, are clearly inferior to those with the simpler LCC of Eqn (4.6), however, and it is the latter which is recommended as an outcome of the present work.

9. Quintic element

In using 4 df/node to this point more accurate interpolation for u is allowed. In the context of accurate calculation of curvatures more accurate interpolation for v is also of interest and hence a quintic interpolation with freedoms v, v_s, v_{ss} at each end is derived by matrix inversion [12] as

$$
\begin{aligned}
f_1 &= 1 - 10s^3 + 15s^4 - 6s^5 \ \ for \ \ v_1 \\
f_2 &= s - 6s^3 + 8s^4 - 3s^5 \ \ for \ \ v_{s1} \\
f_3 &= (s^2 - 3s^3 + 3s^4 - s^5)/2 \ \ for \ \ v_{ss1} \\
f_4 &= 10s^3 - 15s^4 + 6s^5 \ \ for \ \ v_2 \\
f_5 &= -4s^3 + 7s^4 - 3s^5 \ \ for \ \ v_{s2} \\
f_6 &= (s^3 - 2s^4 + s^5)/2 \ \ for \ \ v_{ss2}
\end{aligned}
\tag{9.1}
$$

Using cubic interpolation for u we then have 5 df/node and the results obtained using load stepping and the Cartesian and curvilinear LCCs of Eqns (2.11) and (4.6) are given in Table 6. As in Table 4 the Cartesian LCC results are clearly inferior to the curvilinear ones and it is the curvilinear correction which is recommended.

Generally the 10 df element with the curvilinear LCC is roughly as accurate, but no more so, than the 8 df element and hence the latter is recommended.

Finally, however, a 12 df element with the quintic interpolation of Eqns (8.1) for both u and v was tried. Preliminary results for $e = 4,10$ and 10s/3i were inferior to those with the 8 df element for both LCCs (Eqns (2.11) and (4.6)). The 8 df element gives linear moments and the consistent quadratic strain components in Eqn (6.1) are apparently sufficient for the present work.

TABLE 6. Results for quintic/cubic element

e	s	i	u	v	ϕ
Load stepping:					
4	10	1	-14.4998	47.2871	1.0075
10	10	1	-14.6935	47.2287	1.0080
10	20	1	-15.2645	46.5983	1.0038
10	40	1	-15.5335	46.2765	1.0001
20	10	1	-14.7700	47.2990	1.0094
20	20	1	-15.2692	46.5529	1.0033
20	40	1	-15.5628	46.2671	1.0002
With Cartesian LCC:					
4	10	3	-15.8424	46.2378	1.0112
10	10	3	-16.1281	46.2769	1.0158
10	20	3	-15.8943	46.0596	1.0038
10	40	3	-15.8372	46.0064	1.0010
20	10	3	-16.1842	46.3004	1.0170
20	20	3	-15.9276	46.0552	1.0152
20	40	3	-15.8652	45.9948	1.0010
With curvilinear LCC:					
4	10	3	-15.4327	45.8252	0.9881
10	10	3	-15.6964	45.8560	0.9920
10	20	3	-15.7872	45.9543	0.9980
10	40	3	-15.8105	45.9801	0.9995
20	10	3	-15.7468	45.8759	0.9930
20	20	3	-15.8199	45.9498	0.9982
20	40	3	-15.8383	45.9685	0.9996
Exact:			-15.8529	45.9698	1.0000

10. Stress calculations

To this point stress calculations have not been considered. To this end the thrusts and moments at the integration points in an element are calculated as a summation over the load steps as

$$T = \sum EA B_1 \{\delta d\}_i \tag{10.1}$$

$$M = \sum EI B_2 \{\delta d\}_i \tag{10.2}$$

where B_1 and B_2 are rows one and two of matrix $\overline{B}$ formed as in Section 4 and $\{\delta d\}_i$ are the displacements for load step i, stress calculations being carried out during the last iteration of each load step.

172

TABLE 7. Stress results.

Step, i	Element #	T	M
1	4	0.67	9,917
2	4	1.16	19,835
3	4	1.73	29,753
4	4	2.27	39,671
5	4	2.85	49,588
6	4	3.37	59,506
7	4	3.92	69,424
8	4	4.48	79,341
9	4	4.99	89.259
10	1	0.63	100,007
10	2	-3.72	99,893
10	3	-5.93	99,603
10	4	5.52	99,177
Exact:		0.0	10,000i

The results for 4e/10s/4i are shown for the central integration point (which gave best results) for the fourth element at each load step and for all elements at the last load step in Table 7.

These are satisfactory and for 10e/40s/4i the maximum value of T is only 0.1 whilst the least accurate value of M is 99.924 in the last element and step.

11. Convergence of the LCC

To convincingly demonstrate convergence of the curvilinear curvature correction (Eqns 4.5 and 4.6) with 4 freedoms per node further results were obtained using 25, 50 and 100 elements for the problem of Figure 1. These are summarized in Table 8.

Previous curvature correction results were obtained using single precision MegaBasic computation (8 d.p.). To provide the greater memory needed for larger problems Visual Basic was used to obtain the results of Table 8. Single precision (7 d.p.) gave slight errors for $e = 50$ and serious errors for $e = 100$ so that double precision (14 d.p.) was used for all the results of Table 8.

The results for increasing numbers of load steps are extrapolated by assuming h^2 convergence (where h is proportional to $1/s$) so that the ordinal intercept of the regression line for the plot of v vs $1/s^2$ is the extrapolation result. For all three columns of data $R^2 = 1$ to 5 decimal places was obtained. Accurate regression lines for increasing e can be obtained in the same way.

TABLE 8. Tip deflection v using 3 iterations per load step

Load steps, s	$e = 25$	$e = 50$	$e = 100$
10	45.88721	45.88530	45.88483
20	45.95118	45.94906	45.94853
40	45.96736	45.96514	45.96436
50	45.96931	45.96708	45.96652
100	45.97193	45.96967	45.96910
Extrapolated	45.97270	45.97046	45.96984
Exact		45.96977	

Now applying h^2 extrapolation to the extrapolated results for $e = 50$ and $e = 100$ we obtain

$$v* = 45.96984 + (45.96984 - 45.97046)/3 = 45.96963$$

which is very close the exact solution.

A more practical example of extrapolation is obtained using the results of Tables 2 and 4 for 10 and 20 elements and 20 and 40 load steps. Using h^2 extrapolation for $e = 10$ gives

$$v*_{10} = 45.9829 + (45.9829 - 45.9660)/3 = 45.9885$$

For $e = 20$ one obtains

$$v*_{20} = 45.9690 + (45.9690 - 45.9528)/3 = 45.9749$$

Applying h^2 extrapolation to these two results gives the final result as

$$v* = v*_{20} + (v*_{20} - v*_{10})/3 = 45.9697 \quad [\text{exact} = 45.9698]$$

which is very close to the exact solution. The extrapolation results of Section 6 with only a few elements, however, are probably sufficient for practical purposes.

Finally, applying the foregoing 2-way extrapolation procedure to the results for $e = 50$ and 100 and $s = 50$ and 100 in Table 8 yields $v* = 45.96977$, which is the exact solution to 5 decimal places, convincing proof that the curvilinear LCC is correct.

12. Conclusions

[1] The standard 6 df beam element with the (corrected) Cartesian curvature correction gives satisfactory results, though many load steps are required for good accuracy. The curvilinear correction gives similar results.

[2] Because it uses exact transcendental functions the fictitious load method is more accurate than the two LCC methods. Only the more general LCCs, however, model the actual straining within an element and can be extended to apply to the analysis of plates and shells where the use of the natural strain approach [2, 11, 12] allows the inclusion of such calculations directly.

[3] The 8 df element gives good results when the usual extensional strain correction is included.

[4] The additional extensional strain correction of Eqn (4.3) has negligible effect but may be more significant in rubber-like materials.

[5] With the 8 df element the Cartesian LCC gives modest accuracy but the curvilinear correction is significantly more accurate.

[6] The curvilinear LCC of Eqn (4.6) with the 8 df element is recommended and in Tables 3 and 4 this is shown to converge to the exact solution.

[7] The LCC proposed by Milner is found to work but Eqn (7.2) is incorrect and correct, let alone accurate, results cannot be obtained with this or with a single load step.

[8] Quintic formulations with 5 and 6 df/node were respectively found no more accurate and less accurate than the formulation with 4 df/node.

[9] Stress calculations, summed over load steps, are found satisfactory for the LCC of Eqn (4.6) using 4 df/node.

[10] As the results of Table 4 show, the LCC is elusive in fine meshes with many load steps but it is more clearly advantageous with few elements, as shown in Table 3, and the corrections of Eqn (4.6) (and also (2.11)) are particularly easy to include in computer programs.

[11] The virtual increment calculations of Eqns (2.11) and (4.6) have been verified and such calculations may have wide applications, for example in relativistic gravitational physics [5] where the form of the Lorentz-Fitzgerald contraction can be interpreted as a curvature in space-time [3]. In this context the beam in Fig. 1 is a light streamline and the end loading that of a black hole or sink, causing the familiar swirling or 'plug-hole' effect and hence a curvature in the line.

[12] An objective of future work is to apply the LCC to plate and shell problems using Argyris' natural strain approach [2] so that 'cross derivative' forms of the LCC are not required.

175

References

[1] N.S. Armouti, "Transverse earthquake-induced forces in continuous bridges," *Structural Engng Mechanics* **14** (2002) 733-738.

[2] J.H. Argyris, "Three-dimensional anisotropic and inhomogeneous elastic media matrix analysis for small and large displacements," *Ingenieur-Archiv* **34** (1965) 33-55.

[3] A.S. Eddington, *The Mathematical Theory of Relativity*, 2nd edn (Cambridge University Press, Cambridge, 1924).

[4] H.M. Gomes and A.M. Awruch, "Some aspects on three-dimensional modelling of reinforced concrete structures using the finite element method," *Advances in Engineering Software* **32** (2001) 257-277.

[5] P.C. Kohnke, "Large deflection analysis of frame structures with fictitious forces", *Int. J. Numerical Methods in Engineering* **12** (1978) 1279-1294.

[6] M.M. Lavrentief et al, *Inverse Problems of Mathematical Physics* (Brill/VSP, Leiden, 2003).

[7] H.S. Lee, "Minimum-weight design of a moment-resisting frame accounting for incremental collapse," *Structural Engineering Mechanics* **13** (2002) 35-52.

[8] H.R. Milner, "Accurate finite element analysis of large displacements in skeletal frames", *Computers & Structures* **14** (1981) 205-210.

[9] C.W. Misner, K.S. Thorne and J.A. Wheeler, *Gravitation* (W.H. Freeman, San Francisco, 1973).

[10] G.A. Mohr and H.R. Milner, "Finite element analysis of large displacements in flexural systems", *Computers & Structures* **13** (1981) 533-536.

[11] G.A. Mohr and N.B. Patterson, "A natural numerical differential geometry scheme for a doubly curved shell element", *Computers & Structures* **18** (1984) 433-439.

[12] G.A. Mohr, *Finite Elements for Solids, Fluids, and Optimization* (Oxford University Press, Oxford, 1992).

[13] G.A. Mohr and J.H. Argyris, "The large curvature correction in finite element analysis," *Int J Arts & Sciences* **1** (2001) 1-9.

[14] G.A. Mohr and J.H. Argyris, "The large curvature correction in finite element analysis - II," *Int J Arts & Sciences* **1** (2001) 27-35.

[15] P.J.M. Monteiro, K.P. Chong, J. Larsen-Basse and K. Komvopolous, *Long-Term Durability of Structural Materials* (Elsevier, Oxford, 2001).

[16] S. Utku, "Stiffness matrices for triangular elements of nonzero Gaussian curvature", *J. Amer. Inst. Aeron. Astron.* **10** (1967) 1659-1667.

[17] O.C. Zienkiewicz, *The Finite Element Method*, 3rd edn (McGraw-Hill, New York, 1977).

A.3. Visual BASIC program for LCC calculations

The following VB5/6 coding is that used to obtain the results of Table 8. The coding for the form, which contains a Command button to start calculation is:

```
Private Sub Command1_Click()
Call main
End Sub

Private Sub Form_Load()

End Sub
```

The coding for the element stiffness matrix and loading terms, including some of the large curvature calculations discussed earlier in this appendix is:

```
DefDbl A-H, O-Z: DefInt I-N
Public op As Object
Sub main()
Set op = Form1
Open "\lcc\beam25.txt" For Input As #1
10 Dim EM(8, 8), SM(404, 404), Q(404), NO(105, 2), CO(105, 2), CI(3)
Dim WF(3), HH(2, 8)
12 Dim F1(8), B(2, 8), Z(2), BB(2, 8), T(8, 8), ER(8), D(404), EV(8), G(2, 8), H(2, 8)
13 Dim SE(100), C1(101), S1(101), CD(404), EG(8), F2(8), P(8, 8), ibc(110, 2), qf(30, 5)
15 TF = Sqr(15) / 10: CI(1) = 0.5 - TF: CI(2) = 0.5: CI(3) = 0.5 + TF
17 WF(1) = 5 / 18: WF(2) = 8 / 18: WF(3) = 5 / 18
20 EA = 10: EI = 5: YM = 2000000: D1 = EA * YM: D2 = EI * YM
25 ITL = 3: LS = 10
30 Input #1, np, NE, NB, NL, NBW: NN = 4 * np
op.Print "np =", np
40 For N = 1 To np: Input #1, CO(N, 1), CO(N, 2)
op.Print N; CO(N, 1); CO(N, 2): Next
50 For N = 1 To NE: Input #1, NO(N, 1), NO(N, 2)
op.Print N; NO(N, 1); NO(N, 2): Next
For N = 1 To NB: Input #1, ibc(N, 1), ibc(N, 2): Next
For N = 1 To NL: For j = 1 To 5: Input #1, qf(N, j): Next: Next
52 For NST = 1 To LS
53 For I = 1 To NN: D(I) = 0: Next
55 For ITN = 1 To ITL
57 For I = 1 To NN: For j = 1 To NN: SM(I, j) = 0: Next: Next
60 For N = 1 To NE: NI = NO(N, 1): NJ = NO(N, 2)
67 For I = 1 To 8: ER(I) = 0
68 For j = 1 To 8: EM(I, j) = 0: Next: Next
70 DX = CO(NJ, 1) - CO(NI, 1): DY = CO(NJ, 2) - CO(NI, 2)
EL = Sqr(DX * DX + DY * DY)
80 CA = DX / EL: SA = DY / EL
```

177

```
If ITN = 1 Then
SE(N) = EL: C1(N) = CA: S1(N) = SA
End If
84 For I = 1 To 8: For j = 1 To 8: T(I, j) = 0: P(I, j) = 0: Next: Next
85 P(1, 1) = CA: P(1, 3) = SA: P(3, 1) = -SA: P(3, 3) = CA: P(2, 2) = 1: P(4, 4) = 1
86 P(5, 5) = CA: P(5, 7) = SA: P(7, 5) = -SA: P(7, 7) = CA: P(6, 6) = 1: P(8, 8) = 1
87 CA = C1(N): SA = S1(N): Rem EL=SE(N)
88 T(1, 1) = CA: T(1, 3) = SA: T(3, 1) = -SA: T(3, 3) = CA: T(2, 2) = 1: T(4, 4) = 1
89 T(5, 5) = CA: T(5, 7) = SA: T(7, 5) = -SA: T(7, 7) = CA: T(6, 6) = 1: T(8, 8) = 1
90 For I = 1 To 4: EG(I) = D(4 * NI - 4 + I): EG(I + 4) = D(4 * NJ - 4 + I): Next
95 For I = 1 To 8: EV(I) = 0: For K = 1 To 8
96 EV(I) = EV(I) + T(I, K) * EG(K): Next: Next
100 For II = 1 To 3: S = CI(II): SS = S * S
110 F1(1) = 0: F1(2) = 0: F1(3) = 6 * SS - 6 * S: F1(4) = EL * (3 * SS - 4 * S + 1)
120 F1(5) = 0: F1(6) = 0: F1(7) = 6 * S - 6 * SS: F1(8) = EL * (3 * SS - 2 * S)
125 F2(1) = F1(3): F2(2) = F1(4): F2(3) = 0: F2(4) = 0
127 F2(5) = F1(7): F2(6) = F1(8): F2(7) = 0: F2(8) = 0
130 v1 = 0: For I = 1 To 8: v1 = v1 + F1(I) * EV(I) / EL: Next
132 VG = 0: For I = 1 To 8: VG = VG + F1(I) * EG(I) / EL: Next
135 U1 = 0: For I = 1 To 8: U1 = U1 + F2(I) * EV(I) / EL: Next
140 For I = 1 To 8: BB(1, I) = 0.5 * v1 * (1 - 0 * v1 * v1 / 12) * F1(I)
145 B(1, I) = v1 * (1 - 0 * v1 * v1 / 6) * F1(I): Next
150 B(2, 1) = 0: B(2, 2) = 0: B(2, 3) = (12 * S - 6) / EL: B(2, 4) = 6 * S - 4
160 B(2, 5) = 0: B(2, 6) = 0: B(2, 7) = (6 - 12 * S) / EL: B(2, 8) = 6 * S - 2
163 TF = Sqr(1 - v1 * v1)
165 Rem TF=1+VG*VG:TF=sqr(TF*TF*TF)
166 For I = 1 To 8: BB(2, I) = B(2, I) / TF: Next
167 TF = 1 + v1 * v1 / (1 - v1 * v1)
168 Rem TF=1-3*VG*VG/(1+VG*VG)
170 For I = 1 To 8: B(2, I) = BB(2, I) * TF: Next
180 For I = 1 To 8: B(1, I) = B(1, I) + F2(I) * (1 + U1): Next
182 For I = 1 To 8: BB(1, I) = BB(1, I) + F2(I) * (1 + 0.5 * U1): Next
185 TF = WF(II) / EL
186 For I = 1 To 2: For j = 1 To 8: H(I, j) = 0
187 For K = 1 To 8: H(I, j) = H(I, j) + B(I, K) * T(K, j): Next: Next: Next
190 For I = 1 To 8: G(1, I) = D1 * H(1, I): G(2, I) = D2 * H(2, I): Next
191 For I = 1 To 8: For j = 1 To 8
192 For K = 1 To 2: EM(I, j) = EM(I, j) + H(K, I) * G(K, j) * TF
193 Next: Next: Next
206 For I = 1 To 2: For j = 1 To 8: HH(I, j) = 0: H(I, j) = 0
207 For K = 1 To 8: HH(I, j) = HH(I, j) + BB(I, K) * T(K, j)
208 H(I, j) = H(I, j) + B(I, K) * T(K, j): Next: Next: Next
209 For I = 1 To 8: G(1, I) = D1 * BB(1, I): G(2, I) = D2 * BB(2, I): Next
210 For I = 1 To 2: Z(I) = 0: For K = 1 To 8: Z(I) = Z(I) + G(I, K) * EV(K): Next: Next
212 For I = 1 To 8: For K = 1 To 2: ER(I) = ER(I) + H(K, I) * Z(K) * TF: Next: Next
240 Next II
```

```
245 For I = 1 To 4: Q(4 * NI - 4 + I) = Q(4 * NI - 4 + I) - ER(I)
246 Q(4 * NJ - 4 + I) = Q(4 * NJ - 4 + I) - ER(I + 4): Next
250 For I = 1 To 2: For j = 1 To 2: For IL = 1 To 4
260 IE = 4 * (I - 1) + IL: NR = 4 * NO(N, I) - 4 + IL
270 For JL = 1 To 4: JE = 4 * (j - 1) + JL: NC = 4 * NO(N, j) - 4 + JL
280 SM(NR, NC) = SM(NR, NC) + EM(IE, JE)
290 Next: Next: Next: Next
295 Next N
300 For I = 1 To NB: nf = ibc(I, 1) * 4 - 4 + ibc(I, 2): op.Print "bc df"; nf;
310 For j = 1 To NN: SM(nf, j) = 0: SM(j, nf) = 0: Next
320 SM(nf, nf) = 1: Q(nf) = 0: Next I
330 For I = 1 To NL: N = qf(I, 1): qu = qf(I, 2): qr = qf(I, 3): qv = qf(I, 4): qm = qf(I, 5)
340 Q(4 * N - 3) = Q(4 * N - 3) + qu: Q(4 * N - 2) = Q(4 * N - 2) + qr
350 Q(4 * N - 1) = Q(4 * N - 1) + qv: Q(4 * N) = Q(4 * N) + qm
op.Print "loads"; N; qu; qr; qv; qm: Next
400 For I = 1 To NN
410 X = SM(I, I): Q(I) = Q(I) / X
415 J2 = I + NBW: If J2 > NN Then J2 = NN
420 For j = I + 1 To J2: SM(I, j) = SM(I, j) / X: Next
430 For K = 1 To J2: If K = I Then GoTo 470
440 X = SM(K, I): If X = 0 Then GoTo 470
Q(K) = Q(K) - X * Q(I)
450 For j = I + 1 To J2
460 SM(K, j) = SM(K, j) - X * SM(I, j): Next
470 Next K
480 Next I
482 For I = 1 To np: CO(I, 1) = CO(I, 1) + Q(4 * I - 3)
483 CO(I, 2) = CO(I, 2) + Q(4 * I - 1): Next
485 For I = 1 To NN: D(I) = D(I) + Q(I): CD(I) = CD(I) + Q(I): Q(I) = 0: Next
490 Debug.Print CD(4 * np - 3), CD(4 * np - 2), CD(4 * np - 1), CD(4 * np)
500 Next ITN
510 Next NST
End Sub
```

Line 17 of the Main subroutine gives the weights for the numerical integration (3 point Gauss quadrature) used to obtain the element stiffness matrix in line 192.

Lines 70 to 80 calculate the geometry change for the element.

Following lines up to 182 calculate the B matrix etc. required to calculate the equivalent loads arising from geometry changes, a few Rem lines being used to include alternative LCC correction calculations.

Line 280 assembles the system stiffness matrix.

Lines 400 to 480 a use compact Gauss reduction routine to solve the system equations.

The data file beam25.txt is for 25 elements (and thence 26 nodes):

```
26, 25, 3, 1, 8
0, 0, 4, 0, 8, 0, 12, 0, 16, 0, 20, 0, 24, 0, 28, 0, 32, 0, 36, 0
40, 0, 44, 0, 48, 0, 52, 0, 56, 0, 60, 0, 64, 0, 68, 0, 72, 0, 76, 0
80, 0, 84, 0, 88, 0, 92, 0, 96, 0, 100, 0
1, 2, 2, 3, 3, 4, 4, 5, 5, 6, 6, 7, 7, 8, 8, 9, 9, 10, 10, 11
11, 12, 12, 13, 13, 14, 14, 15, 15, 16, 16, 17, 17, 18, 18, 19, 19, 20, 20, 21
21, 22, 22, 23, 23, 24, 24, 25, 25, 26
1, 1, 1, 3, 1, 4
26, 0, 0, 0, 10000
```

Line 1 of data gives the number of nodes (NP), the number of elements (NE), the number of boundary conditions (NB), the number of loads (NL), and the band width for the problem (NBW = number of nodes/element [2] x number of degrees of freedom/node [4]).

The next three lines give the nodal coordinates CO(N, 1), CO(N,2) for the 26 nodes.

The following three lines give the element node numbers NO(N,1), NO(N,2) for the 25 elements.

The penultimate line of data gives the three boundary conditions with IBC(N,1) being the node number, and IBC(N,2) being the number of the boundary condition with:

1 fixing u = 0
2 fixing du/dx = 0
3 fixing v = 0
4 fixing dv/dx = 0

The last line of data gives the load at the last (26th) node with QF(1) = node number, and QF(1 to 4) giving the degree of freedom, here a loading of 10000 being given for the last degree of freedom i.e. dv/dx, and thus being a moment loading [this in conjunction with, 10 load steps set in line 25 giving the total loading of 100,000 shown in Figure 1].

The number of load steps is set at 10 at line 25 in the program [and the number of iterations per load step is set at 3 in this line], so the foregoing data set (for 25 elements) gives the first result for s = 10 and e = 25 in Table 8. If LS = 20 in line 25, then the last date line above changes to

26, 0, 0, 0, 5000

so that total load = 100,000 as per Figure 1 and the result for s = 20 and e = 25 in Table 8 is obtained.

Appendix B

INTRODUCTION TO BASIC

B.1. A brief history of BASIC

BASIC was developed by Kemeny and Kurtz at Dartmouth College (New Hampshire) in the early 1960s and was much used on minicomputers (which typically had 16 terminals, each being allowed 16 kb of RAM, the amount required by the then versions of BASIC) in the 1970s.

In 1975 the first microcomputer was sold, a clumsy box + switches affair with storage of only 256 bytes. In the same year Tiny BASIC, consisting of just 20 pages of code, was written and many versions of this quickly appeared and, also in 1975, Gates and Allen launched Microsoft Corporation with their version, this being marketed with the Altair microcomputer.

A flood of microcomputers with as little as 16 kb of RAM then appeared, the Apple, the Commodore 64, the Spectravideo, the HP85 and many others, all having their own version of BASIC.

In the early 1980s IBM quit their near monopoly of the electric ('golfball') typewriter market, switching to production of *PCs* with about a MB of RAM. Now there was a flood of PCs: Apple, IBM, ICL, NEC, Olivetti etc., as well as many IBM 'clones.'

On the IBM BASICJ, BASICA and GW ('Gee Whiz') BASIC appeared. All used about 64 kb of RAM and the latter is quite powerful. With the advent of a MB of RAM or more Chris Cochran and American Planning Corp's MegaBasic appeared to make full use of it.

From Microsoft QBASIC, using a rudimentary GUI (graphic user interface), followed and was shipped with DOS 5 whilst Quick BASIC, the first fully compiled BASIC appeared around the same time, and Visual Basic (VB) for Windows shortly thereafter, this having compilation as an option.

VB4 was still somewhat clumsy to use, but VB5 is very user friendly. VB5 is quick, but not as quick as the original computer language, FORTRAN, or the later C++. There are still reminders of its predecessors, for example the QBColor( ) function.

VB6 and VB7 or VB.NET, however, are about as quick as C++ when compiled so that BASIC is finally competitive speedwise.

The original BASIC feature of having a command interpreter allows programs to run on an almost 'line by line basis without full compilation so you don't type the whole program in, compile and receive a long list of cryptic error messages which don't even tell you where the program stopped. Instead mistyped lines produce an immediate error as you type them.

When you do run the program, therefore, there will be only one error message at a time, telling you when the program stopped and you go to that line and correct the error, and thus work your way through what should be only a few errors.

QBASIC, the version of BASIC used in most of the coding given in this book, can be downloaded free from the internet, as can QuickBASIC, a later version which includes a compiler.

In versions of Windows such as Windows XP and Vista, QBASIC must be run in Command Prompt mode.

In later versions of Windows from Windows 7 to Windows 10, QB4.exe can be obtained in 32-bit and 64-bit versions from www.QB64.org by clicking DOWNLOAD under 'QB64 v1.3 out now!' and on the page that then appears, choosing to download either of the zip files: qb64_1.3_win_x64.7z (64 bit) or qb64_1.3_win_x86.7z (32 bit), and also help_1.3.zip.

The qb64 programs convert the QBASIC code to executable C++ with the output from PRINT commands displayed in a separate window.

[QBASIC programs can also be run via a free program which uses a 'DOS box', but this is confusing to use.]

B.2. Introduction to BASIC programming

BASIC commands

The most elementary BASIC commands are:

> RUN - to run a program
> SAVE - to store a program
> ENTER - to add lines
> REN - to renumber program lines (with default 'gaps' of 10)
> LIST - to list the program (on screen)
> BYE - to leave BASIC

Arithmetic operations

The following program determines the square toot of a number using Newton's method in which the root is given by iterating the recursion relation

$$x_{new} = (x_{old} + num/x_{old})/2$$

where num = number for which the square root is required

> x_{old} = initial estimate of the square root

Then, using a tolerance number TOL as a termination criterion the program is:

```
10 Rem SQRT using Newton's method
20 INPUT "Input, xold, num,tol", XOLD, NUM, TOL
30 XNEW = 0.5*(XOLD+NUN/XOLD)
40 DIFF = ABS9XNEW-XOLD)
50 IF DIFF<TOL THEN GOTO  80
60 XOLD = XNEW
70 GOTO 30
80 PRINT "SQRT =", NEW
```

and to test the program typical input is 1,4,0.001 to obtain Ö4 = 2.

Note that ABS( ) is a library function for the absolute value and that in some versions of BASIC a final line, 90 END is required (and in VB a first line Sub MYPROG( ) is needed to declare a subroutine).

In early versions of BASIC line numbers were necessary and in very early versions of BASIC variable names were restricted to two a single alphabetic character plus a single optional digit.

In QBASIC (and VB) line numbers are not necessary and variable names can be many characters but statements are upper case (converted thus if typed otherwise). Then when computation is redirected by a GOTO (or THEN GOTO, for which only half the statement is actually required) statement the target line must have a *label* (e.g. LAB1:) which is given in the GOTO statement. Thus the foregoing example can be written more briefly as

```
INPUT xold, num, tol
LAB1: xnew = (xold + num/xold)/2 : diff = ABS(xnew-xold)
IF diff<tol GOTO LAB2
xold = xnew: GOTO LAB1
LAB2:PRINT xnew
```

where a semicolon is used as a *statement separator*. Line numbers are used in most coding given in the present book, however, in part because they help describe how programs work (i.e. "lines 110 - 160 do - - - ').

Strings

Ease of string handling is one of the traditional advantages of BASIC. The following program reads three names (given in the DATA statements at the end) and prints them (on screen) with three spaces between. It then prints an integer and a real number using PRINT USING to *format* these.

```
10 READ a$, b$, c$
20 x$ = SPACES$(3)
30 PRINT A$;x$;b$;x$;c$
40 P$="#####" : Q$ = "#####.##"
50 n = 2 : c = 14/3
60 PRINT USING P$ ; n ; : PRINT USING Q$;c
70 DATA 'Bob", "Jim", "Ted"
```

Note that a ; follows the 'n' of the first PRINT USING statement to print both numbers on the same line, otherwise the second number will appear on a second line. Here in line 40 P$ is in *integer* format and Q$ is in *real number* format. Strictly variables should be *declared* at the start of the program as *integer, real, double precision* etc.

Arrays and Loops

The following *database* program dimensions (i.e., declares their size) *arrays* and then uses a *loop* (on i) to read some names and ages and print them out, right justifying the names using the LEN function.

```
DIM names$(10),num(10)
FOR i = 1 to 3
READ names$(i), num(i)
j = LEN(names(i)) : x$ = SPACE$( j )
PRINT x$;names$(i),num(i) : NEXT
DATA "Jane" , 25
DATA "June" , 35
DATA "Caroline" , 15
```

Functions

Finally we give a simple example of a user defined function to calculate the square of a number. Note the way the variable X is passed to the function as an *argument* and the function result is returned as R.

```
10 DECLARE FUNCTION SQ(Z)
20 Z=2
30 Y = SQ(Z)
40 PRINT Z
100 FUNCTION SQ(Z)
110 Z=Z*Z
END FUNCTION
```

Note that QBASIC automatically stores the function as a *subroutine* in a separate *workspace* accessed via the VIEW menu from the menu bar (at the top of the screen).

Standard functions

Standard arithmetic, mathematical and string functions used in BASIC include

INT( ) - gives the integer (truncated) value of a number

ABS( ) - gives absolute value of a number (unsigned)

RND(x) - gives a random number [x <0 gives same number, x> 0 (or x not given) gives the next number in the sequence, = 0 gives the last number]

SQR( ) - gives square root

SIN( ) - gives SIN( ) of an angle in radians

LEN(A$) - see example program in "Arrays and loops" earlier in this section

CHR$(n) - gives the ASCII character corresponding to integer n
        (e.g. n = 65 gives A)

Subroutines

The simplest way of forming subroutines is using the GOSUB command to move to program segments appended after the END statement

```
10 PRINT "main"
20 GOSUB 50
30 PRINT "main"
40 END
50 PRINT "sub"
60 RETURN
```

Alternatively subroutines are stored as separate programs and called by a main program. The following program is called MAIN. It has a subroutine 'datin' which is called and numbers passed to it, omitting one number so that it prints as zero when the number list is printed while in the subroutine.

```
DECLARE SUB datin (N, M)
REM MAIN
DIM X(10), Y(10)
COMMON SHARED Y()
X(1) = 5: Y(2) = 3: N = 10: M = 10
datin N, M
PRINT "main", X(1), Y(2), N, M
END

SUB datin (N, M)
DIM X(10)
PRINT "sub", X(1), Y(2), N, M
END SUB
```

Here the argument list passes N, M to the subroutine and the COMMON SHARED statement allows listed variables to be accessed by all other subroutines. As the array X() is not included in the shared statement, X(1) will print from the subroutine as zero.

Data files

Here we give a examples of a data files (as distinct from program files) using the following program.

```
OPEN "c:\basic\temdat" FOR OUTPUT AS #7
OPEN "c:\basic\gmdata" FOR RANDOM AS #8 LEN = 100
x = 2: y = 3
PUT #8, 1, x :PUT #8, 2, y
WRITE #7, x, y
CLOSE #7
OPEN "c:\basic\temdat" FOR INPUT AS #7
GET #8, 2, z : PRINT z
INPUT #7, z : PRINT z
```

185

Here two files are used for *sequential* access and *direct* or *random* access, in the second case overestimating the *record* length and reading back only the second number written to it.

As another example the following code accesses a .DBF file in which a list of names, account numbers, balances and dates is stored:

```
OPEN "\gmwork\accs.dbf" FOR INPUT AS #8
ON ERROR GOTO pend
PRINT "Start"
FOR i = 1 TO 4: INPUT #8, a$
PRINT a$: PRINT: NEXT
pend: PRINT "end"
END
```

The file had data for only three people and was set up using Lotus Approach but .DBF files are used by other programs such as Q&A and Sortit. The ERROR statement is to end the program without error message interruption when end of file (EOF) is encountered. As should be expected, the recovered data includes headings and is printed without formatting. In this instance the account number heading was 'a/c #' which did disturb reading of the headings slightly.

Searching and comparing data

The following code is a very simple example of comparing data, in this case string data. In conjunction with search, therefore, such comparisons can be used to locate specific data.

```
10 a$ = "jim" : b$ = "jim"
30 IF a$ = b$ THEN PRINT "OK"
40 b$="ted"
50 IF a$ = b$ THEN PRINT "OK" ELSE PRINT "NO"
```

In the previous example program such comparisons might be used to extract negative numbers (perhaps corresponding to negative account balances) and the associated personal details from a file.

B.3. Sorting routines

The simplest type of sort is a *bubble sort* in which we successively pass down through the numbers, interchanging pairs of numbers when the second exceeds the first. Eventually the numbers fall into descending order but it takes over 2000 calculations to sort 100 short integer numbers.

More efficient is *search sorting* which seeks out the maximum number of those remaining to be sorted and places this at the top of these. This takes over 400 calculations for the 100 number sort.

More efficient are *hybrid* sorting routines which combine the two approaches and sometimes use *recursion* (i.e., the subroutine calls itself) and take only about 250 calculations for the 100 number sort.

A program using the *Quick Sort* method is given below, this using recursion. It lives up to its name and takes about 180 calculations for the 100 number test.

```
DIM H, L, ii AS LONG
DECLARE SUB quicksort (a(), L, H)
DECLARE SUB partition (L, H, ii, a())
DIM a(101)
FOR i = 1 TO 100: VALUE = RND(.5) * 100: a(i) = INT(VALUE): NEXT
calcs = 0
CALL quicksort(a(), 1, 100)
FOR i = 1 TO 100: PRINT a(i); : NEXT
PRINT "Calcs = ", calcs


SUB partition (L, H, ii, a())
SHARED calcs: DIM i, j AS LONG
piv = a(L): i = L: j = H + 1
REM Choose pivot as first element in range
DO
  DO
   i = i + 1: REM From start look for larger # (if there is)
  LOOP UNTIL a(i) > piv OR i >= H
  DO
   j = j - 1: REM From end look for smaller # (if there is)
  LOOP UNTIL a(j) < piv OR j <= L
          REM If they haven't crossed swap them
  IF i < j THEN
    temp = a(i): a(i) = a(j): a(j) = temp: calcs = calcs + 1
  END IF
LOOP UNTIL j <= i: REM Swap pivot with the split in the array
a(L) = a(j): a(j) = piv: calcs = calcs + 1
ii = j: REM Return index of # in correct location for next 'split sort'
END SUB

SUB quicksort (a(), L, H)
REM If the range is valid then sort
IF L < H THEN
REM Split the array & return index of the item in the correct location
CALL partition(L, H, ii, a())
REM Sort the lower portion
CALL quicksort(a(), L, ii - 1)
REM Sort the upper portion
CALL quicksort(a(), ii + 1, H)
END IF
END SUB
```

Finally note that it is sometimes necessary, and generally a wise precaution, to declare variable types as *integer, real* (the default)*, or double precision* using the DEFINT, DEFSNG, DEFDBL statements. Alternatively this can be done globally as the first line of a program using:

DefSng A-H, O-Z: DefInt I-N

to reserve I-N for integers, as is the default in FORTRAN, the original programming language.

B.4. Visual BASIC (VB)

Visual basic programs usually begin with a form, Form1, for which the coding:

Private Sub Form-Load( )

End Sub

is automatically added when VB starts the new program with Form1.

Then one can add BASIC coding, to obtain, for example:

Private Sub Form-Load( )
Show
x = 2
y = 3
z = x + y
Print "z =", z
End Sub

Here the command Show is necessary to print on the form.
Alternatively, one can have VB add a coding Module to contain the commands of one's program. This will take the form:

Sub main( )
y = 2
x = 3
z = x+ y
Form1.Print "z = ", z
End Sub

and to run this include the command line

Call main

in the coding for Form1.

Regrettably, perhaps, VB does not have DATA statements, so that data must be read from a separate file, though for just a few values of variables, of course, these can be 'declared' as above.

188

As a VB example, the following coding plots simple vibrations.

The coding for Form1, which has a Command button added to it that is pressed to start the program, is

```
Private Sub Command2_Click()
Call main
End Sub

Private Sub Form_Load()

End Sub
```

The coding in the program module for the vibration plotting is:

```
Public output As Object
Sub main()
Rem Time stepping vibration program
Set output = Form1
output.FontSize = 20
output.Print "                VIBRATION PLOT"
output.Line (0, 0)-(0, 0): output.DrawWidth = 5
A1 = 0: B1 = 0: A2 = 0: B2 = 0: XL = 0: SL = 0: F = 40
output.Line (0, 0)-(0, 8000): output.Line (0, 4000)-(8000, 4000)
For T = 0 To 2.5 Step 0.1
S = 0: If T <= 0.5 Then S = 1
A3 = 1.5 * A2 + B2 / 2 - A1: B3 = S + A2 / 2 + B2 - B1: X = 120 * T
Z = 100 - 20 * SL: Y = 100 - 20 * S: X = F * X: Z = F * Z: Y = F * Y
output.Line (XL, Z)-(X, Y)
Z = 100 - 20 * B2: Y = 100 - 20 * B3: Z = F * Z: Y = F * Y
output.Line (XL, Z)-(X, Y)
Z = 100 - 20 * A2: Y = 100 - 20 * A3: Z = F * Z: Y = F * Y
output.Line (XL, Z)-(X, Y)
A1 = A2: B1 = B2: A2 = A3: B2 = B3
XL = X: SL = S: Next
End Sub
```

After running the program the resulting plotting on Form 1 comes out as follows, showing the a small initial disturbance, along with the vibration of two parameters which, for example, might be the movement at two levels of a building subjected to an earthquake.

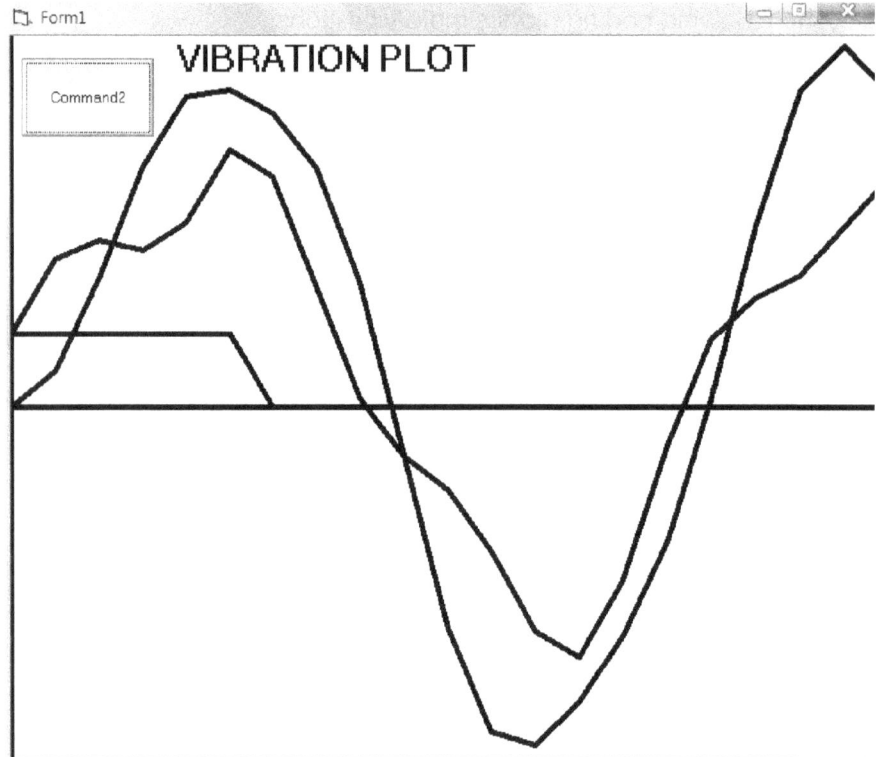

As a further VB example, the following is a VB5/6 listing for the steepest descent program given in Sec. 6.4. A quick search with the four step lengths noted there, and also at the start of this listing, will give a good solution.

The program has a form *Form1* to which is attached a command button. When this is clicked a program module *Module1* is called and an input box appears asking for the value of the penalty factor B. Input 1 and the box asks for search step S. Input trial values and when this search is complete use S = 0 to terminate the search, when a new B value (and then S) is requested.

To terminate execution use S = 0 followed by B = 0.

The coding for Form1 is:

```
Private Sub Command1_Click()
Call main
End Sub

Private Sub Form_Load()

End Sub
```

The coding for routine and subroutine of Module 1 is:

```
Attribute VB_Name = "Module1"
DefSng A-H, M, O-Z: DefInt I-L, N
Private op As Object: Public X1, X2, B, F
Sub main()
Rem searches: B=1, 0.082 & 0.14; B=100, 0.0047 & 0.00035
Set op = Form1: op.DrawWidth = 3: op.FontItalic = True
Dim C(10, 10): I = 0: M = 1.05: S = 0: op.FontSize = 14: op.FontBold = True
op.PSet (2400, 1400), RGB(255, 255, 255): op.Print "Optimum"
op.PSet (1100, 2100), RGB(255, 255, 255): op.Print ">= constraint"
op.PSet (700, 500), RGB(255, 255, 255): op.Print "= constraint"
op.PSet (3200, 1000), RGB(255, 255, 255): op.Print "Solution path"
X = 0: Y = 1000: op.Line (X, Y)-(X, Y)
For Z = 0 To 2 Step 0.01
X = Z: Y = 1 - X * X / 4: X = 2000 * X: Y = 3000 - 2000 * Y
op.Line -(X, Y): Next Z
op.Line (0, 2000)-(4000, 0): op.Line (0, 3000)-(6000, 3000): op.Line (0, 3000)-(0, 0)
X1 = 2: X2 = 2: C(1, 1) = X1: C(1, 2) = X2
X = 2000 * X1: Y = 3000 - 2000 * X2: op.Line (X, Y)-(X, Y)
NEWB: a$ = InputBox("B", , , 5000, 4000): B = CSng(a$)
I = I + 1: C(I, 1) = X1: C(I, 2) = X2: S1 = 0: If B = 0 Then GoTo PEND
Call Subb: Debug.Print "IF =", F
F1 = F: X1 = X1 * M: Call Subb: F2 = F: X1 = X1 / M
G1 = (F2 - F1) / (X1 * (M - 1))
X2 = X2 * M: Call Subb: F2 = F: X2 = X2 / M: G2 = (F2 - F1) / (X2 * (M - 1))
NEWS: a$ = InputBox("S", , , 5000, 4000): S = CSng(a$): If S = 0 Then GoTo NEWB
X1 = X1 + (S1 - S) * G1: X2 = X2 + (S1 - S) * G2
Call Subb: Debug.Print X1; " "; X2; "F= "; F
S1 = S
X = 2000 * X1: Y = 3000 - 2000 * X2: op.Line -(X, Y)
GoTo NEWS
PEND: End Sub

Sub Subb()
G = 1 - X1 * X1 / 4 - X2: E = X1 - 2 * X2 + 1: If G > 0 Then G = 0
FU = (X1 - 2) ^ 2 + (X2 - 1) ^ 2
F = FU + B * G * G + B * E * E
End Sub
```

191

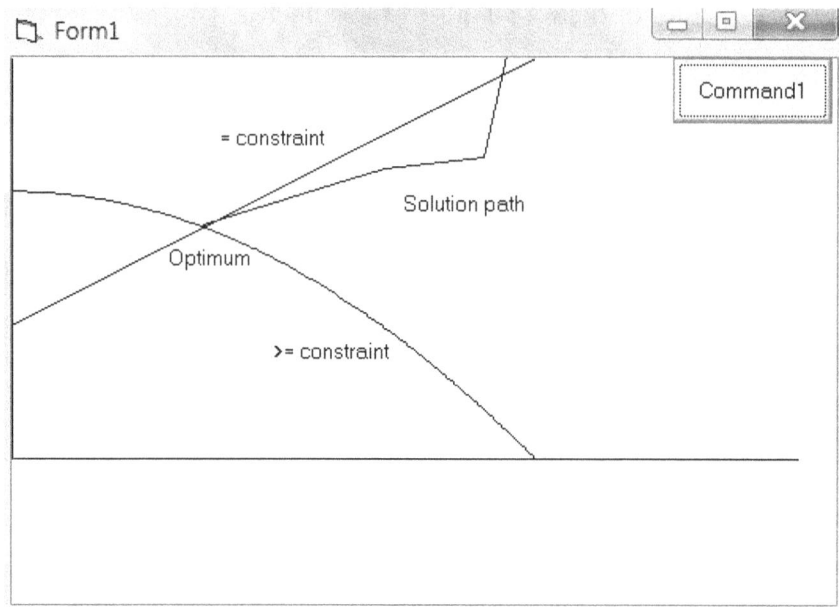

The solution progress plotted on Form 1 with the penalty factors and search lengths stated in the 5th Rem line of Module 1 is shown above.

The main VB differences are that progress results are printed in the immediate window using Debug.Print and plotted on Form1. Generally, however, numerical results would be printed on Form1 and data sets would be read from a file using

Open "\newvb\dsndat.txt" For Input As #8

15 Dim rf(20), cord(20, 2)

20 Input #8, np, ne, ns, nq

Note that in this short example a program module need not be used and the code could have been attached to Form1, this without a command button (or any other control) but with the statement *Show* preceding any executable statements. Then the program is started by the F5 key as for QBASIC.

B.5. Using programs listed in this book

To use/run the programs listed in this book type them into a word processor and save them as ----.txt files. Then rename them as ----.BAS files which can then be read into and run in QBASIC in command prompt mode (or for later versions of windows in QB64, as noted at the end of Section A.1) or as modules in Visual Basic.

B.6. References

Brown S, *Visual Basic in Record Time,* Sybex, Alameda CA 1998.

Capron HL, Williams BK, *Computers and Data Processing*, Benjamin Cummings, Menlo Park CA, 1982.

Cochran C, *MegaBasic Users Manual*, American Planning Corp., Alexandria VA 1984.,

Fox D, *Pure Visual Basic*, Sams 1999.

Kreitzberg CB, Scheidman B, *The Elements of FORTRAN Style,* HBJ, New York, 1972.

Lien DA, *The BASIC Handbook,* 3rd edn, Microtech, Dubai, UAE, 1989.

Perry G, *Introduction to Computer Programming*, SAMS, New York, 2001.

Price WT, *Fundamentals of Computers and Data Processing with Basic*, Holt, Rinehardt and Winston, New York NY, 1983.

Time-Life (eds), *Computer Languages*, Time-Life Inc. 1986.

MS GW-BASIC User's Guide and User's Reference, MS Corp., 1987.

Finite Elements
using Natural Strains & Basis transformation

Key chapters of this important book include:
> Line elements and basis transformation.
> Natural coordinates and strains.
> Thin plate elements using basis transformation.
> Plane stress elements using basis transformation.
> Facet shell elements.
> Thick plate elements.
> Curved shell elements.
> Natural element for potential flow.
> Viscous fluid flow.
> Optimization of finite element models.
> The large curvature correction.
> A useful introduction to BASIC programming.

**G. A. Mohr did his PhD at Churchill College, Cambridge.
He published circa 60 journal papers and 40+ books, including:**

A Microcomputer Introduction to the Finite Element Method
Finite Elements for Solids, Fluids, and Optimization
The Pretentious Persuaders, A Brief History & Science of Mass Persuasion
Curing Cancer & Heart Disease;   The Variant Virus
The Doomsday Calculation, The End Of The Human Race
Heart Disease, Cancer, & Ageing: Proven Neutraceutical & Lifestyle Solutions
2045: A Remote Town Survives Global Holocaust
The History & Psychology of Human Conflict;   The War of the Sexes
Elementary Thinking for the 21st Century
The 8-Week+ Program to Reverse Cardiovascular Disease
The Scientific MBA;   Mohr's Law of Hierarchies
The DIY Cardiovascular Cure;   Combating Cancer
New Ideas for the 21st Century;   Economics: A Basic Introduction
Finite Elements and Optimization for Modern Management
A Half Life: The Memoirs of Geoff Mohr

Also with R.S. Mohr/Richard Sinclair & P.E. Mohr/Edwin Fear:
The Evolving Universe: Relativity, Redshift and Life from Space
World Religions: The History, Psychology, Issues & Truth
World War 3, When & How Will It End?
The Brainwashed, From Consumer Zombies to Islamic Jihad
Human Intelligence, Learning & Behaviour
New Theories of The Universe, Evolution, and Relativity
The Psychology of Hope;   The Population Explosion
Brainwashed Zombies: Religious, Political & Consumer Persuasion
Human Conflict: An Attitudinal Psychology Model